SPLINTERS OF INFINITY

Illustration by Genevieve Leonard.

SPLINTERS OF INFINITY

COSMIC RAYS AND THE CLASH OF TWO NOBEL PRIZE–WINNING SCIENTISTS OVER THE SECRETS OF CREATION

MARK WOLVERTON

THE MIT PRESS CAMBRIDGE, MASSACHUSETTS LONDON, ENGLAND

The MIT Press would like to thank the anonymous peer reviewers who provided comments on drafts of this book. The generous work of academic experts is essential for establishing the authority and quality of our publications. We acknowledge with gratitude the contributions of these otherwise uncredited readers.

This book was set in ITC Stone Serif and Avenir LT Std by Westchester Publishing Services. Printed and bound in the United States of America.

Library of Congress Cataloging-in-Publication Data

Names: Wolverton, Mark, author.
Title: Splinters of infinity : cosmic rays and the clash of two Nobel Prize–winning scientists over the secrets of creation / Mark Wolverton.
Description: Cambridge, Massachusetts : The MIT Press, [2024] | Includes bibliographical references and index.
Identifiers: LCCN 2023019247 (print) | LCCN 2023019248 (ebook) | ISBN 9780262048828 (hardcover) | ISBN 9780262377836 (epub) | ISBN 9780262377829 (pdf)
Subjects: LCSH: Cosmic rays—Research—History | Cosmology. | Millikan, Robert Andrews, 1868–1953. | Compton, Arthur Holly, 1892–1962.
Classification: LCC QC485 .W625 2024 (print) | LCC QC485 (ebook) | DDC 523.01/97223—dc23/eng/20231018
LC record available at https://lccn.loc.gov/2023019247
LC ebook record available at https://lccn.loc.gov/2023019248

10 9 8 7 6 5 4 3 2 1

The supreme social value of science is in the demonstration which it furnishes of the difference between the right and the wrong answer.

—Robert Andrews Millikan, *Cosmic Rays: Three Lectures*, 1939

For those who know science, its inhumaneness is a fiction.

—Arthur Holly Compton, "Science Makes Us Grow," 1941

Jesus Saves
    But Millikan Gets Credit

—Anonymous student graffiti at Caltech, 1937

CONTENTS

PRELUDE: SHAKING HANDS WITH MUSSOLINI

Convegno di Fisica Nucleare, Villa Farnesina, Rome, Italy, October 1931

Wolfgang Pauli was in something of an agitated state. The Austrian scientist found himself in Rome, in a seething mixture of nervousness, excitement, and consternation.

Only thirty-one years old, he was already a rising star in the new and rapidly developing field of quantum physics, working with luminaries such as Niels Bohr in Copenhagen and Max Born in Göttingen, and appointed a professor of theoretical physics at ETH Zurich in 1928. He had already worked out some of the fundamental precepts of the quantum physics revolution, most famously what had become known as the "Pauli exclusion principle," and most recently had come up with a controversial explanation for a nuclear process called beta decay, proposing the existence of a new particle with no mass or electric charge and therefore almost impossible to detect.

Still recovering from a divorce and personal upheaval the previous year, he was in need of some distraction and professional affirmation. He was in Rome for what was billed as the first international conference in nuclear physics, organized by the Royal Academy of Italy, presided over by academy president Guglielmo Marconi and chiefly organized by Enrico Fermi, noted Italian physicist and another leading light of nuclear physics.

"Fermi asked me to talk about my new idea, but I was still cautious and did *not* speak in public . . . only privately," Pauli wrote later in a letter to a friend.[1] Though not known either for shyness or timidity, he was also too careful a scientist to go too far out on a professional limb for the sake of still-developing ideas that might not pan out, and that many other more august colleagues, such as fellow conference attendee Niels Bohr, found decidedly dubious.[2]

So that accounted for the nervousness. The excitement came from the conference itself: the opportunity to meet, socialize, and schmooze with so many fellow scientists and hear their stimulating ideas about the fertile new fields of nuclear physics, quantum mechanics, and the especially controversial topic of cosmic rays. Assembling at the Rome conference was practically a roll call of the world's most eminent physicists, not merely Fermi, Bohr, and Marconi, but also Marie Curie, Werner Heisenberg, Arnold Sommerfeld, Lise Meitner, Peter Debye, and even two American Nobel physics laureates, Robert Millikan and Arthur Compton. The only notable physicist conspicuous for his absence was the world's most famous example, Albert Einstein, who declined to attend, apparently out of protest to the fascist government of Benito Mussolini and its decree earlier this year forcing Italian professors to sign a loyalty pledge to the regime.[3]

This was also the main source of Pauli's consternation, and likewise for quite a few of the other attendees. Under Il Duce's rule, the Royal Academy of Italy had essentially displaced and superseded the *Accademia dei Lincei*, one of Europe's oldest and most venerable academic bodies, which had counted Galileo among its members. Under Mussolini, however, it became largely another organ of the fascist state, ultimately existing to glorify the leaders. Holding prestigious international conferences attended by world-class scholars was an effective way to "promote Italian research and culture in the international peer communities and to give Italy and fascism a good name in international public opinion."[4]

If anyone had any doubts on that score, they had only to note that the opening of the conference on Sunday, October 11, 1932, was presided over by Il Duce, his prime minister, and his other chief lackeys, all resplendent in fascist finery. Willingly or not, Royal Academy of Italy president Marconi publicly praised Il Duce for his support and his ostensible enthusiasm for theoretical physics.

Conference organizer Fermi was certainly no supporter of fascism and would later leave Italy in 1938 when Mussolini followed the example of his fellow dictator Adolf Hitler by enacting anti-Jewish laws. For now, though, he did his best to keep politics out of the conference, assuring participants that the proceedings would remain strictly scientific.

Scientists were supposed to be civil, polite, disinterested, above it all. Still, some concessions had to be made, distasteful as they might be. As Pauli later noted, those mostly came at the formal receptions, as he wrote to his friend: "'*Horribile dictu*, I had to shake hands with Mussolini."[5]

1

"A VERY INTERESTING DISCUSSION"

In the days before routine commercial airline travel, intercontinental journeys weren't casual affairs. Distances, whether across the surface of the globe from one point to another or from the ground up into the reaches of the sky, that would later become inconsequential were anything but casual in the early twentieth century. If you were traveling from the United States to Europe, for example, you first had to get to a major port city such as New York or Boston, which unless you were fortunate enough to live in or near meant a train trip of several days to a week. Then there was the transatlantic voyage, which took one or two weeks, ending at another major port city such as London, and then further travel from there to one's ultimate destination, again usually by train.

If you were an eminent American academic and scientist, that meant putting your lab work and teaching on hold or leaving it in the hopefully capable hands of your graduate students, and clearing your schedule probably for at least a month. So one didn't do such things lightly. You wanted to have a good reason for such an investment of time and energy, not to mention money (even though your university would probably pay for the trip, at least if you were a big shot).

In the United States of the 1920s, Robert Andrews Millikan was definitely a big shot. It would be enough that he was the second American physicist to win the Nobel Prize, but he had also helped to establish the

California Institute of Technology (Caltech), which was steadily becoming a friendly rival of long-established centers of scientific learning such as the Massachusetts Institute of Technology, Harvard University, and the University of Chicago.

To Millikan, taking long and exhausting global journeys all came with the territory of being America's most celebrated and eminent scientist. His 1923 Nobel Prize in physics had come mainly for measuring the charge of the electron, a nagging problem in physics that especially appealed to his passion for exactitude, precise measurement, and hands-on experimentation. It also provided a long-awaited confirmation of his scientific self-worth at last. Until he began the electron work at age forty-one, he'd been leading a respectable but relatively unexciting career, noted mostly as a professor who'd written some successful physics textbooks.

After his Nobel triumph, Millikan turned his talents to the study of the mysterious ambient radiation that scientists had begun to detect in the atmosphere. By 1925, Millikan proved that these mystery "rays" were coming not from the atmosphere or the ground, but from outer space, and dubbed them cosmic rays. Further work by Millikan and others confirmed their extraterrestrial origin.

But as is usually the case with science, solving one mystery merely revealed a completely new one. What exactly *were* these cosmic rays? Specifically, were they charged particles like the electron, or neutral packets of energy such as photons of light? Millikan decided the latter, and given his status and obvious expertise, few were inclined to contest him, at least publicly. The cosmic rays, he declared, were "the birth cries of atoms being born in interstellar space."

Such a statement had more than strictly scientific implications. In those years when even such basic questions as the age and size of the universe, much less its origins, were still far from settled, Millikan was, in effect, proposing an entirely new cosmology, a vision of a universe continually creating and re-creating itself throughout eternity. For the public, such talk of creation inevitably invoked talk of a Creator, especially in the wake of the infamous Scopes trial of 1925. Millikan, who had a decidedly religious bent himself, did nothing to dissuade such associations.

Earlier in 1931, Millikan, never one to shy away from a scientific argument especially if it challenged him personally, had participated in a public debate with the British astronomer Sir James Jeans, then visiting Caltech in Pasadena. Jeans, among other scientists such as Sir Arthur Eddington, rejected the whole "birth cries" notion, holding that the Second Law of Thermodynamics portended an eventual end of the universe as it was finally overtaken by entropy or "heat death." To these scientists, cosmic rays weren't "birth cries" but "death wails" resulting from the annihilation of atoms inside stars.

That was anathema to the scientific and especially the spiritual inclinations of Millikan. A universe that wasn't eternal and unchanging, one doomed to wind down inevitably into cold, dark, ultimate oblivion? *Horribile dictu*, indeed.

Just before Millikan arrived in Rome for the physics conference to (reluctantly) shake hands with Mussolini, he had been in London attending another scientific meeting, which featured another more-or-less civil debate with Jeans, who was now on his home turf. Jeans still contended that cosmic rays proved that the universe was decidedly as mortal as human beings themselves, and was "a mere transitory, ephemeral structure." Millikan shot back that "the evidence that cosmic rays furnish for the annihilation theory is not worth a whoop." His hometown paper, the *Los Angeles Times*, reported that "America's foremost physicist" would return from his European travels "with added fame because of his stirring up of the British scientists by his stout insistence upon his theory."[1]

Millikan came to Rome also somewhat stirred up and cranky, though also looking forward to seeing old friends and colleagues and brainstorming with them over the continuing ferment of new developments in physics. He may have also been expecting a bit of rest and relaxation before moving on to lecture in Paris and elsewhere in Europe, including a visit to Einstein. He was likely not expecting yet another challenge to his stoutly insistent scientific theories.

He was about to be sorely disappointed. Not only were more challengers waiting in the wings, but a younger, not-quite-as-eminent colleague would be inspired by the conference to become his greatest and most persistent challenger of all.

Aside from the fascinating questions being raised and debated in quantum mechanics, nuclear physics, and relativity, the hottest topic in physics in the 1920s and 1930s was cosmic rays. The existence of an enigmatic, seemingly inexplicable form of intensely penetrating radiation coming from somewhere had only been confirmed in 1912, and the work of Millikan and others in the following decade established its extraterrestrial origins. It provided a new scientific frontier of unanswered questions, enormous possibilities, and vast opportunity for a young and ambitious scientist such as Bruno Rossi.

Born in Venice in 1905, he studied physics at the University of Bologna and took his doctorate in 1927. The next year, he took a position at the Physics Institute of the University of Florence in Arcetri, near the house where Galileo had spent his last years and died in 1642. After about a year trying to start up several abortive research projects that might address some fundamental physics question, Rossi came across a paper by a pair of German physicists, Walter Bothe and Werner Kolhörster, that sparked his scientific passion. He had hit upon the subject that would direct the course of his professional life: cosmic rays.

"Until then, I had not been particularly interested in the phenomenon of the 'Höhenstrahlung' or 'cosmic radiation' . . . I had not thought that it would offer, to me at least, a profitable field of research," he remembered. "I had not been seduced by Millikan's well-publicized theory, maintaining that cosmic rays were the 'birth cry of atoms' in cosmic space. . . . To my skeptical mind, this was a romantic idea, lacking sound experimental support. On the other hand, I had accepted, uncritically, the prevailing view that primary cosmic rays were high-energy gamma rays."[2]

The Bothe-Kolhörster paper argued otherwise, describing a series of experiments using a brand-new scientific invention called a Geiger-Müller counter. To Rossi, it "came like a flash of light revealing the existence of an unsuspected world, full of mysteries, which no one had yet begun to explore. It soon became my overwhelming ambition to participate in the exploration."[3]

The paper seemed to indicate that rather than consisting of photons, that is, evanescent rays of electromagnetic radiation such as light or gamma rays, the cosmic rays were actual, physical, separable particles with mass—they were, in other words, "corpuscular."

For Rossi, it was a revelation, and it provided the direction he'd been seeking. Bothe and Kolhörster's work "opened a window upon a new, unknown territory, with unlimited opportunities for exploration . . . thus I began what I yet remember as one of the most meaningful and exhilarating periods of my life."[4] With his handful of graduate students, he immediately embarked on an intensive research program, becoming expert at the new and tricky feat of designing and building Geiger counters, which. as he recalled, "was, at that time, a kind of witchcraft."[5] He soon found ways to improve and build upon Bothe and Kolhörster's experimental techniques.

So it was quite natural that Fermi would invite Rossi, his fellow Italian physicist, to give the introductory speech for the fourth day of the nuclear physics conference, which was to be entirely dedicated to cosmic rays, also known as "the penetrating radiation." As the journal *Nature* later reported, Rossi's speech led to "a very interesting discussion."[6]

Particularly for the sixty-three-year-old Robert Millikan, although probably not in a way he would have preferred. Here was this twenty-six-year-old kid, barely three years out of graduate school, giving an address at such an important scientific conference in front of an audience of distinguished scientists, including Nobelists such as Bohr, Curie, and Marconi, not to mention himself? He could perhaps understand Fermi wanting to give the kid a break, but this was a bit much.

But that wasn't the worst part. After laying out a general introduction to the subject and the current state of research, Rossi moved on to the most pressing question about cosmic rays: were they particles or were they photons? To Rossi, the answer was clear. "The main thrust of this talk was to present what, in my mind, were irrefutable arguments against Millikan's theory of the 'birth cry' of atoms."[7]

Once again, Millikan was being challenged, and this time not by an established and well-respected colleague such as Sir James Jeans, but by some know-it-all stripling. And in front of an august international body of professional colleagues, yet. It was, to say the least, galling.

German physicist and future Nobelist Hans Bethe was one of those colleagues. He later told an interviewer that "Rossi was perhaps the first person who proved Millikan wrong."[8] (He actually wasn't, but such demonstrations rarely happened quite so publicly and suddenly.)

Millikan was far too dignified and composed to immediately overreact. But Rossi knew he had struck a serious nerve. "Such a brash behavior on the part of a mere youngster . . . clearly did not please Millikan who, for a number of years thereafter, chose to ignore my work altogether," he remembered.[9]

But the day was only beginning, and Millikan would have ample opportunity to make his own case. He refrained from anything as unseemly as a direct retort to Rossi or any of his other known detractors. Instead, he strongly reiterated his own positions to the conference audience: "Results of our experiments in cosmic rays seem destined to influence profoundly all theories of the origin and destiny of the universe, not only the present ones but all future theories."

Given the controversies roiling in the cosmic ray field, that seemed a rather innocuous and equivocal statement. But Millikan was only getting started. "Due credit must be given to the fact that opening the door to this amazing new knowledge is owing only to the experimentalists," he said—such as himself, went the implication, rather than the theorists, such as James Jeans. "Theorists always have been numerous, especially in astronomy, and they have placed great confidence in their conclusions."

On the topic of theories, there was that annoying second law of thermodynamics, which seemed to concern so many people. "This has been considered by some as conclusive proof for theories of the origin and destiny of the universe," he noted. "But . . . the energy carried by cosmic rays throughout the universe is as great as, or perhaps even greater than all other radiant energies taken together. No one, however, who speculated on the nature of the universe ever foresaw the existence of cosmic rays. In fact, no one ever dreamed of them."

No one could argue with that statement, at least, and it was perhaps the main reason it had taken about twenty years from the time the phenomenon of a mysterious, ubiquitous form of radiation was first noticed until it was conclusively determined that it was coming from outer space. (The distraction of World War I, which effectively shut down most nonmilitary research for the duration, was another reason.)

As the Associated Press noted, "Experiments conducted the past summer and described to the physicists today, [Millikan] said, showed that cosmic rays came from a point hundreds of millions of miles beyond the sun and visible stars and have nothing to do with either. . . . Cosmic rays originate, Dr. Millikan said, in the building up of atoms from hydrogen at an undetermined point many times the distance of the sun from the earth."[10]

It was, in other words, Millikan's "birth cry" theory yet again. "This building up is continually going on in interstellar space, that is space beyond the visible stars," Millikan explained. "We know that all atoms are built up out of hydrogen."[11]

Perhaps so, some of Millikan's audience may have thought, but that didn't prove that any such "atom-building" process was the source of cosmic rays, or how such a phenomenon might even operate.

The next day of the conference, Millikan enjoyed some additional support from none other than the world's most famous female scientist, Marie Curie, who "stood up like a school teacher before some of the world's greatest physicists."[12] Curie explained how she and her colleague W. Bothe had managed to create an "artificial cosmic ray" by aiming an alpha ray at an atom, producing "a gamma ray of great penetrating power."[13]

Sitting in the audience, Millikan was no doubt smiling. Curie and Bothe's work "supported the previous statements of two Americans, Dr. Robert A. Millikan and Dr. Arthur Compton, that cosmic rays came from the building up of atoms in a field of hydrogen millions of miles beyond the most distant star."[14]

The reporter couldn't resist running the Curie-as-schoolteacher comparison into the ground: "Afterward she answered questions directed at her in German and English as though she were a schoolteacher and the celebrated scientists who asked them were pupils."[15]

The conference ended two days later, by all accounts a smashing success. Much productive discussion and debate had occurred, and despite some of the unavoidable fascist trappings, Fermi and his fellow conference organizers had been able to keep the proceedings free of polemics and politics. They "managed to combine both the necessary firmness in directing the conference with the freedom which is so essential for fruitful discussion," reported *Nature*. As for cosmic rays, "It appears that we are yet a long way from understanding this phenomenon, but a variety of new methods of investigation are now being applied, which at least promise to yield important information."[16]

Robert Millikan could only hope that the persuasive schoolmarm Curie had successfully driven home the lesson at last, and that the cosmic ray battles were finally coming to an end. Instead, he would soon discover that the man who would become his staunchest opponent had found in Rome his inspiration to pick up his sword and shield and take up the struggle in earnest.

Arthur Compton wasn't quite as much of a big shot as Robert Millikan, but he was no slouch either. Compton picked up his own Nobel Prize in 1927 for his pioneering work in x-rays and specifically his discovery of what immediately became known as the Compton effect or Compton scattering, which involves the change in electromagnetic energy occurring when a photon collides with a charged particle.

It proved a seminal discovery that was invaluable for the understanding of both electromagnetic radiation and quantum physics, and most of Compton's research throughout the 1920s continued along similar lines. But by the beginning of the 1930s, he was getting restless for a new direction.

He had already been thinking about cosmic rays before the Rome conference and had done some sporadic work in the subject as far back as 1921. While on a Guggenheim fellowship to lecture in India in 1926, he found himself leading a team of Indian scientists into the Kashmir mountains to conduct cosmic ray measurements. "That was really a memorable expedition," he recalled later, both personally and professionally.[17]

By Rome, Compton was ready. "During the early 1930s, when my x-ray studies were nearing completion, I turned my main attention to the properties of cosmic rays," he wrote.[18] He realized that better understanding of this energetic radiation from outer space could only lead to better understanding of radiation on Earth and deeper insight into myriad questions of nuclear physics.

And now, with the invention of the Geiger-Müller counter, physicists had a powerful new tool at their disposal, to complement those already being used to study particles and radiation such as electroscopes and cloud chambers. With his interest in cosmic rays already intensifying, Rossi's presentation in Rome turned on a bright light in Compton's mind. Perhaps the American *éminence grise* of physics, Robert Millikan, had been hugely annoyed by the young Italian, but "Compton had been highly impressed by Rossi's arguments and by his research," notes historian

1.1 Robert Millikan (far left) chats with Marie Curie at the 1931 Rome conference. Between them, Arthur Compton listens. At the far right is conference organizer Marconi. (AIP Emilio Segrè Visual Archives, Goudsmit Collection)

Luisa Bonolis.[19] Rossi also remembered that his Rome talk "awoke a strong interest on the part of Arthur Compton. . . . He kindly told me, some years later, that my 1931 talk had provided the initial motivation for his research program in cosmic rays."[20]

For Arthur Compton, perhaps more than almost any of the other attendees, the 1931 *Convegno di Fisica Nucleare* of the *Reale Accademia d'Italia*, otherwise known as the Second Volta Conference on Nuclear Physics of the Royal Academy of Italy, was most decidedly "a very interesting discussion." Less than six months later, he would kick off an ambitious global campaign that would take him and a dedicated team of colleagues to the most exotic and untrammeled regions of the planet, from dense jungles and uncharted polar wastelands to the deepest mines and highest mountain peaks, manning specially designed detectors in what was then the largest group of researchers ever assembled on a single project, all to solve the mysteries of cosmic rays once and for all.

As Rossi later wrote, the 1931 conference "marked the beginning of the great debate on the nature of cosmic rays. . . . In the United States the debate was at times bitter, involving the personal prestige of scientists committed to opposing views."[21]

This was true in particular of scientists such as Robert Millikan, who had, as Arthur Compton would soon learn, a strong tendency to take things very personally.

2

THE PENETRATING RADIATION

You're standing alone in an empty field in the dark, nothing but nature around you, the Milky Way and an infinity of stars above. It's utter silence and peace and tranquility.

But not really. Even as you stand there in the dark, you're being deluged by a cacophony of noise and violence from the universe. The peace you're experiencing is deceptive, merely an illusion resulting from the ability of your human senses to apprehend only a tiny sliver of the immensity within which you're immersed.

If you could see or feel them, you would notice that you're being pelted by a subatomic sleet of particles: cosmic rays, pieces of the universe, pieces of creation, coming to and touching earth out of the profoundest depths of space and time.

They are subtle in their presence, detectable only by certain instruments and techniques. The evanescence of their existence is the main reason that human beings have only been aware of their existence for a little over a century. To achieve that awareness, inquisitive individuals had to invent devices that could, even just as a byproduct, demonstrate that something unseen and unexplained was going on.

In the late nineteenth century, physics was just beginning to extend its vision from the visible and the tangible—things that could be seen and touched, held in the hand, things that showed effects that were immediately visible—to the invisible and things too small for the eye to ever perceive. Scientists had telescopes and microscopes and various other devices, but there were always things known to be there that remained tantalizingly out of their range of vision, from the interior of living cells to the far reaches of space.

Among those things was the atom. We had known of its existence for a long time, but what exactly was it? What was it made of? Precisely how did atoms of different materials—iron, salt, wood—differ from one another? Chemists had a fair idea of how various substances behaved and reacted with each other, but the fundamental reasons for what they did remained sketchy.

And what about phenomena such as heat, cold, light, dark, fire, ice, all the various moods of the everyday world? There was so much we didn't know, so many areas out of reach, so many gaps in our knowledge. Most people didn't care all that much. They learned what they had to know to live their everyday lives, and ignored the rest, or filled in the nagging voids with religion, superstition, or whatever convenient nonsense was on offer. Some people, even a certain number of scientists who should have known better, declared that we had already pretty much figured out everything worth figuring out about the physical universe, and all that was left was to work out the numbers to a few more decimal places, dot some i's and cross some t's.

It was true that many of the workings of the physical world were coming into focus. Scientists knew that atoms existed and that different elements were made of different kinds of atoms, although the parts of the atom, whatever they might be, were yet to be discovered and named. Electricity was also a familiar phenomenon, along with the concept of electrical charge and that atoms, or some form of them, could be positive, negative, or neutral. When a normally neutral atom took on a positive or negative charge, it was said to be "ionized."

Then as the twentieth century approached, discoveries started to come thick and fast. In 1895, German scientist Wilhelm Conrad Roentgen was studying the phenomena of fluorescence and phosphorescence, using a special electrical device called a cathode ray tube. Quite by accident, he discovered that the tube generated something more than "cathode" rays, something that could fog photographic plates and leave ghostly images behind, such as the bones inside a hand. He had discovered x-rays.

On the heels of Roentgen's startling discovery, only a couple of months later in 1896, French scientist Henri Becquerel was further investigating Roentgen's work and found another way of messing up photographic film. All one had to do was leave certain types of heavy minerals, such as uranium, on top of a photographic plate and the same thing would occur. It didn't matter if one heavily wrapped up the film in layers of paper or other materials, or even placed it all the way across the room: if uranium was anywhere nearby, the film would become fogged. And this occurred without any cathode-ray tubes or other apparatus in the laboratory. Becquerel realized that some kind of unknown rays were being emitted by the uranium itself. He called it radioactivity.

Soon other radioactive materials were found, by researchers such as Marie and Pierre Curie. It was obvious that there was far more left for scientists to do than just work out a few minor details here and there. As if to prove the point, in 1897 English physicist J. J. (Joseph John) Thomson found that cathode rays, and electric current itself, were composed of streams of miniscule particles. He called them "corpuscles," a term later changed to "electrons." As the century turned, fellow Britisher Ernest Rutherford discovered that radioactivity could be divided into at least two types, alpha and beta, with different properties. Shortly afterward, a third type was identified, gamma radiation, the most powerful and penetrating of all. Rutherford would go on to define the nucleus and with it sketch the first rough picture of the actual structure of the atom.

All these new findings provided some clues to the origins of yet another mysterious, if more subtle, phenomenon that showed up around the same time. One way to detect and store electrical charge was in a device called an electroscope or electrometer (terms that were used interchangeably, in the early days), a simple gadget consisting of two gold leaves hanging from a metal rod inside a small, sealed chamber made of insulating material such as glass. When an electric charge is applied to the rod extending outside the chamber, the gold leaves become charged and move apart from each other, because like charges repel each other. The distance between the leaves shows the amount of charge.

Eventually, however, the gold leaves lose their charge and fall together again. The unanswered question was, why? Some people explained that it was because of random electrically charged dust particles in the air that made the charge leak away. But since air was known to be nonconducting (except in certain circumstances, such as during lightning storms), then what was charging, that is, ionizing, the dust particles in the air? Even if the gold leaves were sealed inside a chamber in which the air was carefully filtered to be dust-free, it made no difference.

The newly discovered phenomenon of radioactivity seemed to be a possibility. Experimenters noted that a radioactive source could steal away an electroscope's charge, because the radiation ionized the air inside. This provided a handy means of detecting and roughly measuring the presence of radioactivity, but it could also be a distinct annoyance in certain types of experiments.

Therefore, reasoned some, the fact that electroscopes lost their charge over time even in the absence of known radioactive sources had to mean that the scopes were somehow picking up radiation from elsewhere, such as the Earth itself. One way to check this theory was to completely shield the electroscope with materials that were known to absorb radiation, such as lead. It would be like sitting in a completely dark, windowless room. It might be the sunniest day ever outside, but you would never know it, because you would be immersed in the deepest darkness.

Rutherford and other scientists tried the theory out in various permutations, using many kinds of materials. It didn't work. They found that whatever caused the ionization inside the electroscope could be reduced, depending on the particular material used and its thickness, but not eliminated entirely, even if the electroscope was entirely surrounded by lead walls thicker than a bank safe. The only possibility was some kind of extremely penetrating radiation, which meant either x-rays or gamma rays, coming from somewhere. But where?

If the answer couldn't be found in the laboratory, then perhaps it could be tracked down elsewhere. If the "penetrating radiation," as it was now more or less officially known, was coming from the Earth's crust in some way, then it was logical that it would diminish with greater distance from the ground, such as high in the air, or increase deeper inside the Earth, such as inside a mine or underwater in a lake or ocean.

That was work not for lab-rat scientists but those of a more adventurous stripe. "The physicists who carried out these experiments had to be great sportsmen at times," noted French researcher Pierre Auger. "As occasion demanded, they became divers, mountaineers, miners, or airmen."[1] Fortunately there was no lack of such daring types, and throughout the early years of the twentieth century, a number of researchers carried their electroscopes to the rooftops of tall buildings, into salt mines and railway tunnels buried inside mountains, and even suspended them underwater from boats. Different researchers came up with different results, likely because they were using different techniques and sometimes operating under rather challenging conditions. Also, the only instrument at their disposal, the electrometer, wasn't exactly the most sophisticated, accurate, reliable, or easily transportable scientific device for measurements in the field.

It took the German researcher Theodor Wulf, a Jesuit priest teaching physics in the Netherlands who had become quite interested in radioactivity and the "penetrating radiation," to improve the situation somewhat. He designed a new and improved electroscope that used thin, metal-coated quartz strings in place of the traditional gold leaves. Wulf's instrument was precision-built by a respected German instrument firm to exacting standards to be easily portable, rugged, and highly resistant to all the various and fickle environmental conditions it would encounter in the field. It essentially became the gold standard for electroscopes for some time thereafter, with a basic design still in use today.

Father Wulf took his instrument to all the various locales that had become de rigueur for electroscope readings: caves, mines, rooftops, under lakes and rivers. Like his colleagues, he concluded that the source of the penetrating radiation was mostly elements in the Earth's surface, with some variations depending on the absorption qualities of the particular materials intervening between the source and the electroscope, air pressures and temperatures, and similar factors. A small amount of the radiation might also be coming from the atmosphere as well, he concluded.

But he was a restless sort and wanted to gather yet more data. In the spring of 1910, on a visit to Paris, he decided to take his electroscope for measurements to the highest place available: the top of the Eiffel Tower, about 900 feet above the ground. Instead of a sharp reduction in the penetrating radiation at that height compared to the ground because of the greater volume of air absorbing it, as would be expected if it were coming from the surface, the ionization was only somewhat less. Either part of the radiation had to be coming from another source such as the atmosphere above, or the air between the surface of the Earth and his electroscope absorbed far less of it than he had calculated.

The only way to be sure one way or the other was to take measurements at far greater altitudes. Back then, airplanes were a brand new and largely unproven technology, and certainly incapable of venturing very far above the ground. That left balloons, which meant braving the vagaries of finicky weather and winds in sometimes rickety and not always controllable craft to get electroscopes as high as possible into the sky. Again, not a job for scientists most comfortable working in warm, safe laboratories.

Ballooning was definitely a far more established enterprise in the early 1900s than venturing aloft in winged, powered airplanes, and was widely used for both scientific and military endeavors. But investigating the penetrating radiation would require making series of measurements at various heights up to the greatest altitudes possible, to look for and record any changes in the phenomenon. That posed an entire slew of daunting problems that typical balloonists at lower altitudes didn't have to consider, including intense cold that froze equipment, scientific instruments, and human beings, as well as variations in humidity and atmospheric pressure that could throw off measurements and make them essentially useless. Also, of course, was the annoying fact that the higher one went, the thinner the air and the harder it was to breathe.

A handful of intrepid souls nonetheless embraced the challenges and took to the air. Among them was Swiss physicist Albert Gockel, who made a few flights with a Wulf electroscope beginning in December 1909, his highest venture to about 14,000 feet. His results were also confounding. The penetrating radiation didn't appear to diminish as the conventional wisdom would have it, but neither did it seem to increase with altitude; instead, it remained more or less constant.

Things were clearly more complicated than previously thought, and in some manner that eluded easy resolution. Victor Hess, who was not only a sharp young physicist from Austria but also something of a fiend for the exotic sport of ballooning, decided that this whole penetrating radiation issue would provide the perfect opportunity to combine his passions for both experimental physics and for floating high above the Earth in a gondola.

He began by conducting careful studies of the response of an improved Wulf electrometer to gamma rays from radium samples, measuring what was called their absorption coefficient in air. This allowed him to precisely calibrate his own instruments with the results obtained by Wulf, Gockel, and others, and to confirm that their findings were more or less accurate. There was no question: the mystery radiation was not being absorbed as it should be by ever-greater thicknesses of air.

Hess started probing the skies with two flights in 1911, using balloons kindly provided by the Royal Imperial Austrian Aeronautical Club. Along for the ride was a pilot named Oberleutnant Heller, who would keep them aloft while Hess tended to the three separate Wulf electroscopes that he brought along, one of them with thinner walls to detect beta rays. One flight was in the daytime, the other in the nighttime, to see whether time of day and the presence of the sun in the sky made any difference. (Note that flying in a balloon at night, in early twentieth-century Austria in total and abject darkness with almost zero artificial light from the Earth below, was hardly an outing for the timid, although Hess thought it had some advantages because the air and winds tended to be relatively calmer then.)

Hess's initial results were frustratingly inconclusive and comparable to those of his airborne predecessors. He attributed this mostly to the relatively low altitudes of these first flights, no higher than about 3,500 feet. But even so, the persistence of the penetrating radiation made him think that it couldn't all be simply earthbound. "The inference was drawn that, in addition to the gamma radiation of radioactive substances in the earth's crust, there must exist still another source of penetrating radiation," Hess wrote later.[2]

He knew he had to keep going, higher, higher. And there was more to be done at lower altitudes: "There was a need for longer flights at low altitudes to carry out measurements and thereby obtain reliable mean values."[3] In 1912 Hess managed to score official funding for seven more flights from the government, in the form of the Imperial Academy of Sciences in Vienna, the *Kaiserliche Akademie der Wissenschaften,* which considered supporting such projects a boon for not only scientific reasons but also nationalistic ones as far as the Austro-Hungarian Empire was concerned.

Hess's voyages would be carried out with scrupulous scientific technique. Before he and his two-man crew of a pilot and a weather observer took flight, Hess made careful control measurements on the ground with his three electroscopes, improved to even better precision based on his previous experiments, to rule out any interference from ground radiation (such as from the sandbags used for ballast).

The flights began in April 1912 and continued into the summer, again sometimes at day, sometimes at night, once during a near-total solar eclipse. Once more, they seemed to confirm the earlier results: no significant decrease of the radiation at higher altitudes, no relation to the time of day. But for reasons of wind and weather and ballooning unpredictabilities, none of the first six flights managed to get any higher than just under 7,000 feet.

Early in the morning of August 7, 1912, just after 6:00 a.m., Hess and his crew lifted off from Aussig on the River Elbe, in a balloon filled not with coal gas as the previous balloons had been, but with hydrogen, in the hope for higher altitude. For the next six hours or so they drifted, with Hess taking measurements at regular intervals of time and altitude. The hydrogen worked: their balloon, named "Böhmen," achieved a maximum height of about 17,500 feet around 10 o'clock that morning.

But most importantly to Hess, he had verified for the first time that when one got high enough, the mystery radiation actually began to *intensify* continuously, instead of diminishing or just staying about the same.

All other reasonable explanations had been systematically ruled out. "If one restricts oneself to the point of view that only the well-known radioactive substances in the earth's crust and in the atmosphere emit radiation with the character of gamma rays and produce an ionization in a closed container, then great difficulties face any explanation," Hess wrote. The results of his flights and those of previous researchers "may most easily be explained on the assumption that a radiation of very great penetrating power impinges on our atmosphere from above."[4] He called it *Höhenstrahlung*, "radiation from above."

The following year, a German scientist and fellow aeronautical daredevil named Werner Kolhörster proceeded to confirm Hess's work by taking five more balloon flights to what were for 1913 quite insanely risky altitudes, up to around 30,000 feet, or almost six miles. He found that at that dizzying height, the *Höhenstrahlung* was about fifty times greater than on the ground. To anyone paying attention, such findings pretty much abolished any notions that the main source of the odd "penetrating radiation" had much to do with minerals buried in the Earth's crust.

2.1 Victor Hess returns to Earth after his historic flight on August 7, 1912. (Wikimedia, American Physical Society)

As it happened, however, Hess wasn't actually the first person to come up with the notion that this radiation, or at least a large part of it, might be coming from somewhere "above." Even before Hess began his probings above the surface of the Earth, Italian scientist Domenico Pacini was probing below it, mostly underwater in oceans and lakes, using Wulf electroscopes. Others had submerged their instruments under water, of course, but Pacini did more than that: he not only went somewhat deeper, but also farther out to sea on military vessels, reasoning that if the radiation was coming from rocks in the Earth's crust, then getting farther from land should produce a measurable difference. He found none.

"The observations made in the ocean in 1910 led me to conclude that a significant part of the penetrating radiation found in air had an independent origin from . . . the terrestrial crust," he wrote in a paper. Even when his instruments were out in the ocean, whether on the surface or underneath the waves, there remained an ionization source "of such intensity that could not be explained by considering the known distribution of radioactive substances in the water and in air."[5] The source, he speculated, could be something in the atmosphere, or even above it—in other words, from outer space.

Unfortunately for Pacini, his work remained rather obscure, and probably even unknown to Hess and the other ballooning adventurers until much later. The reasons why are open to speculation. The robust international network of scientific communication and publishing and conferences had not yet been fully established, for one thing, so news of startling new research discoveries and breakthroughs usually traveled at a fairly stately pace. Other factors worked against Pacini as well. "Why wasn't Pacini's work properly recognized?" Per Carlson asked in a *Physics Today* article. "He carried out his experiments alone and under conditions made difficult by lack of resources. He was, for instance, unable to attend international conferences. And the fact that most of his articles were written in Italian probably contributed to their neglect."[6] Some have argued that Pacini, rather than Hess, should be given primacy as the "discoverer" of cosmic rays.

All this work, however, didn't provide any solid clues as to what the *Höhenstrahlung* or penetrating radiation or whatever one chose to call it actually *was*. It was clearly intensely powerful and penetrating in nature, and the only known type of radiation of comparable nature was gamma rays. "Since gamma rays have the largest penetrating power of the three known ionizing radiations, it was natural to assume that also cosmic rays consist of energetic gamma rays," noted physicist Michael Walter.[7] That particular assumption would eventually cause a major amount of contention and controversy that would spill over from strictly academic and scientific circles into a public debate that would involve everyone from reporters to clergymen.

But that would take a few years as yet. For now, the problem for Hess and his colleagues was that nobody was really paying much attention to their aeronautical and scientific derring-do. If Domenico Pacini was being unfairly ignored, then so were many of his colleagues. Although a small, select clique of scientists found all of this inordinately fascinating and exciting, they had hardly set the scientific world on fire. Not when there were far more important things happening in physics, such as, for example, Albert Einstein's work on relativity and the photoelectric effect, or even the discovery of x-rays, which had obvious and immediate practical implications and applications. But mysterious rays from somewhere out in space? Who cares?

"The discovery of the 'Höhenstrahlung' remained almost unnoticed," wrote Walter. "Probably nobody of the small number of scientists working in this field were sure that the measurements were correct." Aside from that problem, "There was a group of physicists who had serious doubts that a new radiation of cosmic origin was discovered. Their main arguments ranged from a possible radiation in the upper atmosphere to measurement problems due to insulation leaks caused by the low temperatures at high altitudes."[8]

Observed researcher George R. Steber, "Things got so bad that for a time it was not even fitting to speak about the subject of this mysterious radiation, as it seemed more like witchcraft than physics."[9]

At the time, it seemed, the scientific achievements were almost incidental. Yataro Sekido noted in a history of early cosmic ray research, "That wonderful balloon flight by Hess turned out to be one of the last notable scientific achievements of the Austro-Hungarian Empire. Only two years were to elapse before the Archduke Franz Ferdinand, heir to the Habsburg throne, would be killed at Sarajevo."[10]

That event, of course, would trigger World War I. As a byproduct, it also quickly shut down almost all scientific research for the next several years, except the type that had some direct and practical military application. Scientists from nearly all of the combatant nations found themselves, willingly or not, sucked up into the war effort, something for which physical scientists, such as experimental chemists and physicists, were particularly useful. While the chemists worked on projects like devising ever-more-devilish poison gases and explosives, the physicists worked on improving military vehicles and artillery pieces. Some of them became involved in a problem that didn't even exist in any previous human conflict: the detection and destruction of submarines.

One of these scientists was Robert Millikan. By the time his country joined the war in 1917, he was already well established as a prominent member of the American scientific community and a full professor at the University of Chicago. He was known among his colleagues as rather stubborn, egotistical, arrogant at times, extremely sure of himself (at least outwardly)—but also unquestionably a brilliant scientist, teacher, and administrator. He was also not a fan of many of the revolutionary ideas that seemed to be taking over the theory and practice of physics in the early twentieth century, such as relativity and quantum theory. "Millikan was a conservative in the midst of a revolutionary world," noted his biographer, Robert Kargon.[11]

Millikan was also ardently patriotic, passionately spiritual, and never one to shy away from a good argument. All of those qualities, and others besides, were soon to find new and unexpected avenues of expression. Some of the groundwork was being laid already during his war service. He was a member of the National Research Council, intimately involved in organizing the scientific response to preparing and then running the U.S. war research effort. First, he was head of a subcommittee for devising methods of submarine detection (a paramount concern ever since the sinking of the *Lusitania* in 1915), then was asked to actually join the army as a commissioned officer in the Signal Corps, putting his scientific skills and knowledge to use in practical military applications.

One of the many projects he became involved with was meteorology and the use of balloons for various purposes, not only for weather forecasting but also to carry propaganda leaflets over enemy lines. Although Millikan likely didn't realize it at the time, a familiarity with ballooning would eventually pay off handsomely in his scientific career.

Although Hess, Gockel, and a few others managed to carry on a few more balloon trips and mountaineering expeditions here and there during the war years, nothing substantial came of it aside from reiterating and refining what was already known—and what remained unknown. Like most other areas of science in the early years of the twentieth century, however, the investigations of the penetrating radiation were almost exclusively a European affair, with no Americans involved.

As the world, especially war-ravaged Europe, began to heal after the armistice and as the 1920s arrived, that began to change. The leader of the U.S. efforts in experimental science, including what would soon be known as "cosmic rays," would be Robert Millikan.

3

A ONENESS ABOUT THE WHOLE OF NATURE

For Americans at least, "to the public of the 1920s, Millikan represented science," wrote Robert Kargon in his biography of Robert Millikan.[1] If some other aspirants to that status might be around at the end of World War I, no one could reasonably question it after 1923, when Millikan became the second American physicist to win the Nobel Prize. The Nobel conferred an exalted status to any individual, automatically imbuing an added importance and gravitas to any and every statement, opinion, judgment, or speculation that person might care to make, whether or not it had anything at all to do with the subject for which the Prize had been awarded.

The most prominent example of that phenomenon, of course, was Albert Einstein, particularly after a 1919 total solar eclipse provided clear and unmistakable evidence that confirmed his general relativity theory. His ideas about space and time and gravitation might be counterintuitive and incomprehensible to the average person, not to mention some scientists, but the apparent bending of starlight by the Sun's gravity that Einstein had predicted could be observed and measured. No complicated mathematics required, just a good telescope.

After Einstein received the Nobel Prize in 1921, his status as the world's greatest sage and Ultimate Authority on Everything was cemented in the popular imagination all over the world, even if the man himself never claimed to be anything more than a scientist, trying his best to do his work and understand the universe.

Before Einstein's canonization by the world into a secular scientific saint was complete, however, Robert Millikan wasn't quite convinced by his work. After wrapping up his seminal electron experiments before the war, Millikan set himself the task of testing the validity of one of the major achievements of Einstein's 1905 *annus mirabilus* (miracle year), his theory of the photoelectric effect. "He sought to brake what he was to call Einstein's 'unthinkable,' 'bold,' and 'reckless' hypothesis of 'an electromagnetic light corpuscle of energy *hv*.' . . . His research program was an essay in refutation. He saw himself as the defender of the wave theory of light," observed Kargon.[2]

And not incidentally, also a defender of his own electron work, which was coming into question from some quarters and to which Einstein's ideas might also pose a challenge. "In Millikan's eyes, the measurement of the electron charge was to be his legacy to posterity, a legacy that he subsequently jealously guarded," Kargon observed. "He viewed it as his monument, which was not to be easily overturned."[3]

Millikan would later come to accept, if somewhat grudgingly, that Einstein's work was essentially correct. But throughout his career, he was always quick to respond to any professional challenge. "His identity was wrapped up in his achievements," wrote Kargon. "Attacks on his science were converted by him into assaults on his self."[4] And when he believed he was right about something, it took quite a lot to convince him otherwise.

The international scientific community was still in a considerably agitated, restless state in the aftermath of World War I. Bad feelings and hostile attitudes persisted among the former combatants, with much of the resentment directed at citizens of the former Austro-Hungarian Empire. Unfortunately, that also included many of the most important physicists, who happened to hail from Germany and other nations formerly part of Kaiser Wilhelm's domain, and who found themselves either informally or formally ostracized and excluded from publishing in prestigious journals and attending certain conferences.

Those tensions faded and eventually disappeared into insignificance, but it was a slow and painful process. For example, it wasn't until as late as 1928, fully ten years after the end of the war, that a delegation of German mathematicians was welcomed at an international conference in Bologna, Italy. Their leader, David Hilbert, remarked on the "long, hard time" that was finally ending, noting that "it is a complete misunderstanding of our science to construct differences according to peoples and races, and the reasons for which this has been done are very shabby ones. Mathematics knows no races . . . for mathematics, the whole cultural world is a single country."[5]

Still, scientific work and research went on as peace reigned anew, including in the realm of the *Höhenstrahlung*. While some European researchers picked up where they had left off before the war, hauling electroscopes to mountaintops and into the sky, the Americans were now joining the party, led by Robert Millikan, who was yet to be convinced that the explanation wasn't earthbound.

But there were ways to find out. "I had discussed in 1914 and '15 these experiments in our seminar at Chicago, and expressed the opinion that the field seemed to be one of intense interest which should be attacked by discarding manned balloon flights and sending suitable instruments to much higher altitudes than could even begin to be attained in manned flights," he wrote in his autobiography. Unmanned balloons eliminated the problems of keeping a crew warm and breathing. Along with that, he recalled, "my wartime responsibilities in the field of meteorology had whetted my appetite for high-altitude work."[6]

In 1922, he requested funds from the Carnegie Institution of Washington (today known as the Carnegie Institution for Science) for, among other things, a program of sounding balloon flights carrying new instruments that could take readings automatically. "That was a revolution: no more men carried together with instrumentation by a big balloon, but small and cheap unmanned balloons barely carrying the instrumentation," said Mario Bertolotti.[7] Of course, while unmanned flights were less expensive and safer, they "took away also the flavour of the adventure."[8]

For Millikan, noted Kargon, it was an opportunity "once again to bring clarity and precision to a scientific problem area. . . . Millikan saw his problem as first to demonstrate what had merely been indicated, that existence of the extra-atmospheric penetrating radiation, and second, if it existed at all, to determine its characteristics."[9]

Millikan got his money and proceeded to put together his program, with physicist Ira Sprague Bowen as assistant. Again, the basic Wulf electroscope provided a solid foundation for an instrument design built to withstand the rigors of the highest altitudes that Millikan's sounding balloons could achieve. His wartime experience paid off in other ways as well: always skilled at politicking, Millikan had built up an impressive coterie of friends and allies in the military who were all too willing to help him in securing access to equipment and facilities, such as the Army Air Service's Kelly Field outside San Antonio, Texas, which would serve as a staging area. Other Millikan pals in the Signal Corps would help to track the balloons in their journeys.

By this time, Millikan had left the University of Chicago to become one of the founders and head of the physics department at the new California Institute of Technology in Pasadena, California, better known as Caltech. Although he had been well established in Chicago and had spent most of his professional life there, his colleague, the eminent astronomer George Ellery Hale (most noted for establishing the Palomar Observatory), wooed him away after a long campaign to become the chairman of the Executive Council, Caltech's president in function if not official title.

As such, Millikan was eager to help establish Caltech as a fresh new center of American science on the West Coast that could hold its own in prestige and accomplishments with its older brethren back East. His new research program would be a means of possibly doing just that, by exploring another cutting edge of scientific inquiry.

Millikan's balloons finally began lifting off from the Texas hardscrabble terrain in the winter of 1921–1922. Along with the specially designed electroscope equipped with a camera setup to photograph and record the readings, they carried a barometer and thermometer to monitor the environmental conditions in flight, so that none of the necessary parameters would be missed.

The experimental design was brilliant in its simplicity. Each flight would consist not of a single balloon but of two identical balloons, each eighteen inches in diameter, just enough to lift the instrument package, which Millikan's shrewd design had limited to only about seven ounces—less than half a pound. At some point in the ascent, one of the balloons would inevitably burst, and the remaining balloon would ease the instruments gently back to earth.

That left the issue of recovering the instrument package, without which all would be for naught. Aside from allowing the expensive and meticulously constructed instruments to be reused, whatever valuable data they had obtained would be lost if they couldn't be recovered. If the balloon came down in government-controlled territory, there would be no problem, but balloons had an annoying tendency to be unpredictable and to drift wherever the winds might take them. What if Millikan's balloons happened to alight on some remote cattle ranch somewhere?

But Millikan had provided for that eventuality as well. Each package contained a note explaining to whatever curious rancher had found it that the odd gizmo was an important scientific experiment and offering a cash reward for the prompt return of the intact and untampered-with device. That proved an effective strategy, one that Millikan would frequently employ on many future balloon flights in the coming decades.

"The work of Millikan's team made history not only because of the scientific results obtained but also because of the novel and ingenious techniques employed," Bruno Rossi wrote later.[10] Ingenuity notwithstanding, however, any hopes Millikan may have had that the Kelly Field balloon flights would settle the outstanding questions about the penetrating radiation were dashed.

Over four flights, with the highest reaching an altitude of just over nine and a half miles, Millikan and Bowen found that the radiation definitely increased with altitude, as Hess and Kolhörster had observed. But then there seemed to be an actual decrease, at least according to one measurement, followed by an increase. Such results, at best, were inconclusive, at least as far as any "radiation from above" was concerned. Perhaps, Millikan reasoned, the source was somewhere in the high atmosphere itself, someplace more local than outer space. That would seem to be confirmed by the fact that their measurements in general seemed to be far lower over Texas, only about 25 percent of what had been recorded by researchers flying over central Europe.[11]

Millikan and one of his graduate students, Russell Otis, supplemented the Texas balloon flights over the next year with further high-altitude measurements, this time made not from balloons but from airplanes flown from several Southern California airports, and by lugging instruments atop Mount Whitney in California and Pike's Peak in Colorado. The results were similarly ambiguous. They noted in a *Physical Review* paper that, at least judging from the data collected thus far, "there exists no such penetrating radiation as we have assumed,"[12] and that "the whole of the penetrating radiation is of local origin."[13]

But Millikan was too good and careful a scientist to make unequivocal and definitive pronouncements too early. He knew that more experiments and more data would be required and started to figure out what to do next.

If things had gone somewhat differently, Robert Millikan might have ended up not as an American physicist and educator, but as a gym teacher.

That may almost have been fitting, given his origins, which were probably as quintessentially and solidly, unpretentiously American as it was possible to be. He was born on March 22, 1868, in Morrison, Illinois, to Congregationalist minister Silas Millikan and his wife Mary Jane Andrews Millikan, the very essence of Midwestern pioneer Protestant stock. Silas's work took the family across the nearby Mississippi River to Iowa in 1873, first to the town of McGregor and then to Maquoketa, where Robert, the second-born of six children, would grow up.

As his biographer Robert Kargon notes, most of what we know of Millikan's early years comes from Millikan's own recollections in his autobiography and other writings, but by all accounts, it seemed fairly typical for the times: an upbringing that served to instill a strong belief in hard work, self-sufficiency, and solid values. "He worked hard, saved money, was punctual, and was diligent in all things," wrote Kargon. "His report card from Maquoketa High School in September 1886 was splendid: 'Robert Millikan,' it said, 'gained an excellent standing for good deportment, faithful application to duty and exceptionally high grade of scholarship.'"[14]

Like both of his parents, Millikan went on to Oberlin College in Ohio, studying the classics and becoming adept in Greek and Latin, along with some science and mathematics courses required of all the undergraduates. Encouraged by a physics professor, he soon discovered a strong interest in and talent for physics and mathematics, and even ended up teaching an elementary physics course while still an undergraduate.

He was more than simply a nerdy bookworm, though, also active in various athletic pursuits, especially boxing (perhaps predictive of his later, more intellectual, professional combativeness). So he was more than a little surprised when, on the basis of recommendations submitted without his knowledge by some of his Oberlin professors, he was awarded a graduate fellowship in physics at Columbia University. He gave up any ideas of becoming a physical education instructor and instead went on to become Columbia's first Ph.D. in physics in 1895.

After postgraduate work in Germany, Millikan was offered a position as assistant professor of physics at the newly founded University of Chicago by the man who was then America's premier physicist, Albert Michelson. Millikan had worked with Michelson during a summer off from Columbia, so he was already familiar with both his new boss, who had become his scientist idol, and his new institution and new city.

He began to find himself disillusioned after a little while, however, weighed down with teaching responsibilities that left him little time for research—the main reason he'd decided on going to Chicago in the face of other more lucrative offers. He was gaining an impressive reputation as a teacher and author of textbooks, but now he was pushing forty with no significant experimental work or publications to his name. His personal life took a decided upswing when he married the former Greta Blanchard in April 1902, followed by a seven-month honeymoon traveling around Europe.

That not only gave Millikan the chance to renew old professional acquaintances and forge some new ones, but also to get a closer sense of all the groundbreaking work going on in physics all over the Continent. He was fascinated by radioactivity, the yet-unknown workings of the atom, and what secrets those workings held about how elements were created. By the end of 1907, he had settled on the research question that he would, by diligence and persistence, make his own: What was the precise charge of the newly discovered electron?

Robert Kargon noted a couple of reasons for Millikan's choice. First, "he was insistent on research that would address itself to matters of central concern to . . . research problems at what he and his professional colleagues perceived to be the frontier . . . the charge of the electron, *e*, was a fundamental quantity for contemporary physical theory; without a precise knowledge of it, modern physics was a house built on sand."[15]

The other reason was perhaps less professional and more personal. "Millikan wanted to do work that would command the respect of Michelson and his Chicago colleagues, and earn the attention of the world's leading physicists as well."[16] Pinning down the elusive electron, then, would be a way for Millikan to both make a fundamental scientific contribution and prove his worth as a scientist to his idol and to himself. "It may well be that Michelson's receipt of the Nobel Prize in 1907, for measuring the precise speed of light, also influenced Millikan," wrote Judith R. Goodstein.[17]

Millikan's series of electron experiments, in which he ingeniously used oil drops instead of the rapidly evaporating water drops used in previous attempts to measure charge, has been widely and intricately described elsewhere, beginning with Millikan's own exhaustive book on the research,[18] and remains justly famous, not only as a model of elegant experimental design and precise results but also, on a more personal level, his own dogged persistence.

"To get the data needed on one particular droplet sometimes took hours," he recalled in his autobiography. "One night Mrs. Millikan and I had invited guests to dinner. When six o'clock came I was only half through with the needed data on a particular drop. So I had Mrs. Millikan apprised by phone that 'I had watched an ion for an hour and a half and had to finish the job,' but asked her to please go ahead with dinner without me. The guests later complimented me on my domesticity because what they said Mrs. Millikan had told them was that Mr. Millikan had 'washed and ironed for an hour and a half and had to finish the job.'"[19] According to one account, one of the guests, a faculty wife, later said to her husband, "I know we don't pay our assistant professors very much, but I didn't think they had to wash and iron!"[20]

Millikan did succeed in his goal, defining the electron charge to the value still used today, although improved techniques since then have allowed greater precision than was possible in his time. Some have also questioned his work and accused him of "data cooking" in the form of cherry-picking his results, though such accusations have never been convincingly substantiated; the fact that many others repeated his experiments using his techniques and obtained the same results speaks eloquently for their legitimacy.

He also succeeded in his other main objective. His reputation as a scientist was now made, his preeminence as an experimentalist fully established. When the Royal Swedish Academy of Sciences bestowed the Nobel Prize upon his broad Midwestern American shoulders in 1923, his triumph was complete.

Time to move on to the next challenge. This time, it would involve not merely solving another important outstanding question in physics, but just possibly also explaining the nature and origins of the very universe itself.

When Millikan applied to the Carnegie Institution for research funding in 1922, he wasn't just thinking about the penetrating radiation. He had in mind nothing less than "a joint attack on the constitution of matter, the nature of radiation, and the genesis of the elements from the standpoint of physics, astrophysics, chemistry, and mathematics."[21]

He certainly couldn't be accused of harboring modest ambitions. What he, along with his Caltech colleagues George Ellery Hale and Alfred Noyes, was proposing amounted to a wide-ranging research campaign encompassing nearly all the outstanding contemporary questions in physics. It would be a three-part program, utilizing all of the already-existing resources of the new California Institute of Technology and helping to create new ones along the way, not incidentally also putting Caltech on the map as a world-class research center.

For Millikan, pinning down the nature of the penetrating radiation would also provide a way inside the atom itself, and perhaps reveal how the elements of the universe were being created. With all the recent discoveries of radioactivity, the nucleus of the atom, x-rays, and mass spectrometry, it was clear that there was a lot going on inside the atom, and that much of the phenomena that had come to light may all be interrelated somehow. One possibility was called "atom building."

A hundred years earlier, a British chemist named William Prout had advanced the idea that hydrogen, the lightest element, might in fact be the fundamental piece that in various combinations made up all the other heavier elements, whose atomic weights were all simply multiples of hydrogen. "Prout's hypothesis," as it became known, turned out to be far too simple to explain the atoms, but the idea had an appealing quality that persisted, later inspiring Ernest Rutherford when he was busily bombarding atoms with alpha particles a few decades later.

Using a technique called mass spectrometry, Francis William Aston, another British scientist, succeeded in discovering different isotopes of various elements, coming up with a new twist on Prout's hypotheses called the "whole number rule." The rule states that except for hydrogen, the atomic weights of all the elements are whole numbers.

If radioactivity involved atoms breaking down in some manner, then it seemed possible that things could also work the other way around—that new atoms, elements, could be built up from simpler ones. It could be what used to be called "alchemy," the transmutation of matter from one type to another, such as lead into gold. It could be the very secret of creation itself.

Millikan saw in this idea a definite analogy between physics and biological evolution. If living things could arise from simpler forms, as Darwin had proved, why couldn't nonliving things do likewise? Perhaps here was the explanation for everything, the ultimate secret of the universe.

It was an idea that he found irresistibly compelling, and one that would drive him personally and professionally for most of his life. As he would later proclaim in one of his many public lectures, "There is an interrelatedness, a unity, a oneness about the whole of nature, and yet still an amazing mystery."[22] Such a comforting, all-encompassing oneness appealed to the highly spiritual minister's son on the deepest levels.

If the penetrating radiation *was* coming from outer space, it might be the signature of atom building, the energy left over when light atoms combined into heavier ones. That possibility had spurred his original prewar interest in the subject, although he remained uncertain that the ultimate source of the rays was actually outside of the earth. The Texas balloon flights and the mountain jaunts had been frustrating and inconclusive. Something was missing, something was being overlooked.

In August 1925, this time working with G. Harvey Cameron, Millikan tried a somewhat different approach to resolving the issue. They took their electroscopes to a couple of lakes high in the Southern California mountains. One, Muir Lake, was about 12,000 feet above sea level, near Mount Whitney, and the other, Arrowhead Lake in the San Bernardino Mountains, at about 5,000 feet. Both lakes were chosen because they were snow-fed, their water coming from the melting icepack of the mountaintops, and therefore of greater purity and free of any possible residual radioactive contamination from the immediate geological surroundings.

Millikan and Cameron sank their specially waterproofed instruments to various depths within each of the cold, pristine lakes, carefully measuring ionization rates. Beginning at Muir, they found that although the readings went down with increasing depth as would be expected, the ionization persisted even as far down as over eighty feet. That indicated a very penetrating radiation indeed.

The key element was the 6,700 feet difference in altitude between the two lakes. That meant that there were 6,700 more feet of air at Arrowhead

3.1 Millikan (left) and G. Harvey Cameron with their electroscopes and other equipment, circa 1925. (AIP Emilio Segrè Visual Archives)

compared to Muir, with an absorption coefficient equivalent to about six feet of water. What would happen, then, if they took readings and allowed for this six-foot difference between the higher and lower lakes?

What they found astonished them. "Within the limits of observational error, every reading in Arrowhead Lake corresponded to a reading six feet farther down in Muir Lake, thus showing that the rays do come in definitely from above, and that their origin is entirely outside the layer of atmosphere between the levels of the two lakes," Millikan and Cameron reported in *Physical Review*.[23] Along with the data from their previous balloon and mountain experiments, that also seemed to rule out another theory about the penetrating radiation that some had been suggesting, namely that it originated somehow from thunderstorms. No, it wasn't coming from the Earth in any way, neither the rocks nor the air. It definitely was *Höhenstrahlung*, "radiation from above"—"above" meaning outer space.

Millikan was now convinced scientifically, and also pleased personally. Here, it seemed, was confirmation that the facts of science appeared to mesh beautifully with his spiritual leanings. Reporting on his work at the National Academy of Sciences meeting on November 9, 1925, in Madison, Wisconsin, he came up with a better name for the newly confirmed phenomenon.

Rather than something as inelegant and imprecise as "the penetrating radiation" or something as difficult and indecipherable to the English-speaking public as *Höhenstrahlung*, his term had more than a bit of romance and poetry to it, evoking the profoundest mysteries of eternal time and space: "cosmic rays."

4

THE DIN OF BATTLE

Even if it had demonstrated nothing else of any significance, Millikan's public and complete acceptance of the cosmic nature of the penetrating radiation proved one thing: the man could certainly change his mind.

But it took the weight of a huge amount of evidence and argument to do so, because once he had become convinced of something, he hung onto it with the most dogged tenacity, more than willing to take on anyone presumptuous enough to challenge him. Millikan was soon to adopt what would become probably the most prominent idée fixe of his professional, and to a large extent his personal, life.

As physicist Harvey Lemon would later write in his 1936 survey of the cosmic ray field, "Millikan's contagious enthusiasm, his fertile imagination, his ever sanguine assurance of the superior accuracy of his qualitative measurements and the correctness of his interpretations, together with the rather dogmatic style in which he phrases his writings, at once fills any field of study which he enters with a host of other workers, and soon resounds the din of battle!"[1]

The next field of battle, although this particular one would only amount to a minor skirmish, was over the "cosmic ray" term. Millikan's National Academy of Sciences presentation in Madison had been closely covered by the press. That would be routine for the public pronouncements of an

American Nobel laureate, but this time was special. "The National Academy address excited public attention as little else had previously done in American science," noted Robert Kargon.[2]

As only one example, a week after the conference came a *New York Times* editorial that, even given the often purple journalistic prose of the time period, depicts Millikan and his work in almost religious terms.

Dr. R. A. Millikan has gone out beyond our highest atmosphere in search for the cause of a radiation mysteriously disturbing the electroscopes of the physicists," it began. "His patient adventuring observations through twenty years have at last been rewarded. He has brought back to earth a bit more of truth to add to what we knew about the universe. There is no human satisfaction that can be greater than adding a fragment to the body of ascertained truth. He found wild rays more powerful and penetrating than any that have been domesticated or terrestrialized, traveling toward the earth with the speed of light and yet of almost unimaginably short wavelengths . . . here are these till now unknown and even now mysterious forces playing in the great spaces between our earth and the moon—forces of whose origin we know no more than we do of the origin of life on the earth itself.[3]

Given such fulsome musings, the average reader of the *New York Times* with only a passing awareness of science could be forgiven for thinking that this great explorer of truth and the universe, Robert Millikan, had actually discovered these "wild rays," rather than non-American scientists such as Victor Hess, Werner Kolhörster, and their colleagues who, unlike Millikan, had actually risked their own necks making their "adventuring observations" in the cold thin air miles above the ground.

Still, the *New York Times* wasn't done. "The mere discovery of these rays is a triumph of the human mind that should be acclaimed among the capital events of these days," continued the editors. "The proposal that they should bear the name of their discoverer is one upon which his brother-scientists should insist." Instead of something like "penetrating rays," said the *Times*, "they would more appropriately bear the name of the penetrating mind that passed through the miles of space to the far frontiers of our atmosphere and there met these strange forces of the universe coming from space—the mind that lived among them for years to learn their ways, and at last brought us word of their mysterious existence."[4]

To the *New York Times* editorial board, then, there was only one obvious and fitting name for the phenomenon. "'Millikan rays' ought to find a place in our planetary scientific directory, all the more because they would be associated with a man of such fine and modest personality."[5]

The head of even the finest and most modest individual who had ever existed would no doubt be turned by such praise. However, although modesty was not a personality trait for which Robert Millikan was particularly noted, he apparently didn't comment one way or the other on the term "Millikan rays," and certainly didn't encourage the term, continuing to call them cosmic rays. That became the generally accepted term to this day, although *Höhenstrahlung* and its alternative form *Ultrastrahlung* hung on in most European scientific circles for a while longer.

In fact, although he is generally credited as the originator of the term by most sources, some, particularly in Europe, contend otherwise, such as German physicist Michael Walter, who stated: "Today it is assumed that Millikan created the terms 'cosmic radiation' and 'cosmic rays.' . . . But this can also be disputed. Gockel and Wulf used it in a paper from 1908."[6]

"This is an interesting example for Millikan's 'abilities' in publicity and science marketing," said Walter. "Millikan's aggressive campaign had a strong impact."[7] At least in the American press not only at the time

4.1 Millikan "detecting the cosmic pulse" on the cover of *TIME* magazine, 1927. (*TIME*)

but persisting for years thereafter, Millikan was often referred to as "the discoverer of the cosmic ray." To the public, disinclined to make fine distinctions of scientific primacy and who really did what and when, that was good enough, along with the fact that Millikan was unquestionably American.

When the American press reports and publication of Millikan's work reached Europe, Victor Hess was decidedly not amused. He responded in the German scientific journal *Physikalische Zeitschrift* claiming that his own earlier work and that of his colleagues, such as Kolhörster and Gockel, had already established the fact of cosmic rays, while also criticizing Millikan's experiments and techniques. The term "Millikan rays," therefore, "must be rejected as equivocal and unjustified." In his own response, Kolhörster said, "Millikan tries to construct as great as possible an opposition to me so that his results are not immediately recognized as a confirmation of my previous ones."[8]

A somewhat chastened Millikan responded, defending his work and also pointing out that, despite the effusive coverage of the American press, he hadn't originated or supported the "Millikan rays" term or claimed to have actually discovered cosmic rays. He pointed out in a personal letter to Hess that "the really important thing is that between all of us we have been able to make pretty certain the existence of a radiation which comes to earth from outside," which was previously not completely accepted by some other scientists, Millikan included.[9]

Another unpleasant undercurrent to this particular discussion was the bad feelings between German scientists and the rest of the world that hung on after World War I. "The controversy must be placed within the context of the difficult international relations in science at the time, which no doubt affected the feeling and behavior of the protagonists," wrote historians Michelangelo De Maria and Arturo Russo. "By 1925 and 1926, the German speaking scientific community was slowly recovering from the ostracism it had suffered after World War I."[10] With all that in mind, it would be quite understandable that German scientists would be more than slightly upset by having their hard-won achievements apparently being professionally usurped by an American, Nobel laureate or not. (Hess would ultimately be vindicated in 1936 when he was awarded the Nobel Prize for the discovery of cosmic—not Millikan—rays.)

Kargon observed that "Millikan's embarrassment, while genuine, was undoubtedly mixed with considerable pride and satisfaction in the public acclaim."[11] He had scored another unquestioned triumph for American science. Now it was time to continue onward and build upon his achievements to attain ever loftier levels.

With the extraterrestrial nature of cosmic rays proven and almost universally accepted, more equally nagging questions remained. We finally knew, at least in general, where cosmic rays were coming from, true. But what *were* they, and what caused them?

For most physicists, their original name, "penetrating radiation," seemed to be the major clue. It was already known that some forms of radioactivity, namely alpha and beta rays, were composed of different types of atomic particles. In other words, they were physical, if miniscule, objects, or in a term that was popular at the time, "corpuscular," meaning they were made of corpuscles, like blood cells.

But x-rays and gamma rays were something else entirely. They weren't particles but photons, packets of energy that behaved like waves in some instances and particles in others, in ways that the new quantum physics was beginning to explain. Because they were photons, electromagnetic radiation just like light or radio waves, they had wavelength and frequency—but the shortest wavelengths and highest frequencies, which is what gave them their incredibly penetrating qualities. And the cosmic rays, as the measurements of Millikan and others had demonstrated, were able to pierce through thickness of lead as though it wasn't even there, far more easily than any other radiation known.

Over the next couple of years after the Madison announcement, Millikan and Cameron confirmed and refined their work with further measurements in California lakes and also farther afield in the Bolivian Andes, with other readings taken during the ocean trip to South America. With all this new data along with what they had previously collected, Millikan began to work out more precise computations of cosmic ray energies based on their absorption. As historian Peter Galison described it: "The strategy for measuring cosmic ray energy was this: Plot the rate of ionization measured in the electroscope against the absorber thickness. Then try fitting exponential curves with different slopes to the measured one.

The curve with the best-fitting absorption constant would then reveal the photons' energy."[12]

Millikan's travels were duly reported in the press. The *New York Times* noted that his South American sojourn "brought interesting results, confirming the existence of the cosmic ray," and that they weren't comparable with the cathode ray experiments by others that had been in the recent news. "The cathode rays are corpuscular while the cosmic rays are ethereal, [Millikan] said. 'The former consist of particles while the latter are high frequency light,' he added, 'you might as sensibly compare an elephant and a radish.'"[13]

Millikan sat down to analyze all the data he and Cameron had gathered through the lens of current knowledge, including Einstein's mass-energy equivalence, Aston's atomic mass measurements, and a formula developed by British physicist Paul Dirac for determining photon energies lost through absorption. He came up with something that, for him, seemed nothing less than a revelation.

In a February 1928 letter, he told his son Glenn of some new results "which gave me quite a fever." His computations with Cameron demonstrated that the cosmic ray absorption curve could be defined as the sum of three separate energy bands, each corresponding with the extra energy that would be released as hydrogen atoms came together to form the three more complex elements of helium, oxygen, and silicon—"the three elements which constitute the great bulk of the mass of the earth, of meteorites and of the stars." If the results held up, Millikan continued, it would be "the first evidence that the building up as well as the disintegrating process is going on under our eyes, the signal of the birth of an atom of helium, oxygen or silicon being sent out to the ends of the universe wherever such an event occurs in the obstetrical wards of space."[14]

It seemed to Millikan that he had worked out a profound truth, he had unlocked a cosmic secret of all eternity. The excitement he expressed to his son is touching: "Maybe it is a group of accidental occurrences, but I doubt it!!!" [15]

He began to call it his "atom building" theory, or to describe in a more picturesque turn of phrase that cosmic rays amounted to the "birth cries" of new elements. The magazine *Popular Mechanics* offered an early example in June 1928: "The rays, he declares, are 'announcements sent through the ether of the birth of new elements.'"[16] There would be much talk of birth cries and creation coming from "the Chief" at Caltech for at least the next decade.[17]

Later in 1928, a refinement of Dirac's method for deriving energies from absorption curves called the Klein-Nishina formula was introduced, and Millikan was able to fit his data in agreement, providing him with more confidence that he was right. Other scientists, such as the eminent Ernest Rutherford in England, warned against being too hasty, observing that even if the Dirac and Klein-Nishina formulas held fast at relatively lower energies, that didn't necessarily mean that those calculations could be smoothly extrapolated into far higher energy ranges such as those seen in cosmic rays.

Such objections failed to shake Millikan's confidence, however, because for him there was another deeper and more personal significance operating. The atom-building theory, wrote Robert Kargon, was for Millikan "a striking confirmation of his faith in an evolving atomic structure, that is, one in which atoms were being constructed as well as radioactively decaying. It was a way to avoid the 'heat death' or running down of the universe posed by the second law of thermodynamics. . . . In short, atom building was an idea in deep harmony with Millikan's fundamental spiritual yearnings."[18]

Peter Galison noted likewise: "For Millikan, the importance of the discovery of radioactive decay was that it strongly suggested that the inverse process was occurring somewhere in the universe, preventing the much discussed thermodynamic 'heat death' in which the universe would end up cold, maximally entropic, and lifeless. . . . Millikan's broad methodological presuppositions had direct consequences for his experimental physics."[19]

At first, Millikan was at least publicly somewhat circumspect about making sweeping pronouncements about his ideas. The *New York Times* reported in March 1928 that while Millikan had announced, "Discovery of evidence tending to show that the process of creation is now going on in the heavens and that the earth, instead of being a disintegrating world, as has long been believed, is a continuously changing and evolving one," he also said, "It must be taken with some reserve and must be subjected to further critical analysis and further experimental tests." That, however, didn't stop him from saying that "we may have some confidence in the conclusion that the heretofore mysterious cosmic rays, which unceasingly shoot through space in all directions, are the announcements sent out through the ether of the birth of the elements."[20]

In a more detailed exegesis a few days later, *New York Times* science editor Waldemar Kaempffert described Einstein's theory of mass-energy equivalence and how it provided the foundation of Millikan's ideas: "When hydrogen is transformed into helium, something must happen to 0.03 parts of the mass. What? If matter disappears, what are we to expect instead? Look for energy, says Einstein. That 0.03 is converted into radiation—much of it into x-rays of the highly penetrating cosmic type. On this all astrophysicists are agreed. And now Professor Millikan comes forward with experimental proof that his cosmic rays are born when new matter is created in the stars—when hydrogen is transmuted into helium and heavier atoms. And what is his proof? A spectrograph with lines in it . . . in the very positions in which they should be if the Einstein theory is correct—if mass is actually radiated away in the form of energy. In the whole range of physics it would be difficult to find a more satisfying demonstration of the truth of the great principle which Einstein developed mathematically."[21]

Kaempffert concluded, "Millikan has made one of the striking, fundamental discoveries of our day. Physicists were sure that elements were being transmuted in the stars. He proves their case for them. . . . The main thing is that Millikan has removed the doubt once entertained as to the origin of the cosmic X-rays. They are unquestionably created when new matter is born somewhere in the heavens."[22]

As later events would eventually reveal, a considerable amount of doubt actually remained, not only at the time but for long afterward. For the moment, though, the press was happy to celebrate another Millikan triumph. Following New Year's Day 1929, the *Los Angeles Times* published a lavish spread describing how Southern California had become a major center of learning, giving Caltech and Robert Millikan a large part of the credit: "From the experiments with cosmic rays now being conducted by Millikan and Cameron, a knowledge of the universe is now being gained which, in the words of Millikan, 'may help man to live in the future at least a million times more wisely than he now lives.'"[23]

Never shy about self-promotion, Millikan used the opportunity to push atom building and the birth cry theory. "'Dr. Cameron and I,' he continues, 'have recently found three definite cosmic-ray bands or frequencies of penetrating powers, or ray-energies. . . . This discovery of a banded structure in cosmic rays shows definitely that these rays are not produced as x-rays. We have arrived at fairly definite evidence that the observed cosmic rays are signals broadcast throughout the heavens of the births of the common elements out of positive and negative electrons. . . . Observed cosmic rays are, in fact, the birth cries of the infant atoms of helium, oxygen and silicon.'" As if that weren't enough, he added, "We have some little indications that we can also hear the birth-squeaks of infant iron, but we are not yet ready to definitely assert it."[24]

Especially in the hometown papers, some of Millikan's press coverage didn't pretend to any kind of scientific explanations but consisted merely of fawning encomiums, such as columnist Lee Shippey's profile of him. "Some men are world-famed because they have flown higher, run faster or made funnier faces than anyone else," it began. "Dr. Robert Andrews Millikan is famous merely because he is one of the greatest scientists in the world. His discoveries have solved problems over which scientists had disputed for a century."[25] The article studiously documented Millikan's humble beginnings and subsequent life with quotations, apparently from an interview with Millikan himself, such as "I inherited a certain conscientiousness and tenacity from my Scotch-English ancestors," and that man was now living in "the greatest age the world has known or can know for a long time to come."

Millikan, said Shippey, "is white-haired but vigorous, alert, fond of swimming, golf and tennis. . . . He has three children and in social and family life throws off scientific broodings and shows a charming personal side. But when he gets absorbed in a scientific problem he forgets time, fatigue, hunger, everything."[26]

"One more step in unraveling the mystery of the creation of matter was reported to the National Academy of Sciences today by Dr. Robert A. Millikan, Nobel prize winner and director of the California Institute of Technology," announced an Associated Press report on Millikan's 1929 NAS presentation. "The evidence presented was of a form of radiation called cosmic rays. . . . Dr. Millikan explained why he thinks they represent the creation of matter." He did not, however, "assert that he had proved this creation, but said it is indicated, and the only apparent stage from which it can emanate is the vast cold reaches among the stars."[27] Not from the stars themselves, however, as some other scientists maintained.

There were some impressive names among those other scientists, beginning with Rutherford and other notable scientists such as James Chadwick, who thought Millikan's data could have other interpretations, and astrophysicists such as Arthur Eddington and James Jeans. Although Eddington had advanced the idea that stars generate energy by fusing hydrogen into helium and perhaps heavier elements, it didn't follow that cosmic rays were involved. Instead, according to Jeans and Eddington, cosmic rays could be the result of atomic particles annihilating each other deep inside the hearts of stars—death rattles rather than "birth cries." But thoughts of cosmic annihilation and universal heat death were clearly far less appealing than talk of creation and endless rebirth in an eternal universe.

By the late 1920s, cosmic rays were beginning to firmly establish themselves in the public zeitgeist, showing up in advertisements, comic strips, radio plays, and even in religious tomes, a development that probably pleased Millikan if he noticed it. "Persons who enjoy first-class preaching or the moralistic interpretation of important events and great literature will appreciate 'The Cosmic Ray in Literature' by Dr. Lewis Thurber Guild," stated a *Los Angeles Times* column titled "Sermons on Literature." Described as "one of California's most eloquent ministers" who had penned "a fitting appreciation of Dr. Robert A. Millikan's discovery of the cosmic ray,"[28] Guild's book "mentions his brother ministers as 'men of the burning heart who constantly fling that cosmic ray which lighteth every man into the commercial dust, the moral vapors, the psychic fog, the pregnant chaos.'"[29] What any of this might have to do with the details of cosmic ray physics was left undescribed.

Another uncredited *Los Angeles Times* commentary on a supposed "sacred shrine" in a Malden, Massachusetts graveyard in which "stories of miraculous cures" had been reported proceeded to expound the virtues of "a faith that understands," while also managing to invoke the numinous appeal of the newly fashionable cosmic ray: "We should acknowledge as large a field for the unfolding of faith as we willingly afford the miracles of science. And to the uninstructed the law of relativity, the motion of the stars, the induction of the electric current, the composition of the atom, the cosmic ray, are as wholly mysterious as the cure of bodily ailments by the power of faith alone."[30]

The subject of cosmic rays was also noted early on in what seems a more likely venue, the new publishing genre of pulp science-fiction magazines, which viewed it in a different manner. In the September 1929 issue of *Science Wonder Stories*, publisher Hugo Gernsback (who had also founded the first dedicated science fiction magazine, *Amazing Stories*, three years earlier), spoke of "hidden sources of power" in an editorial: "One for instance, that probably will become a tremendous factor in the future, the so-called cosmic ray, was recently discovered by Professor Milliken [sic]. We know practically nothing about it as yet, but from the little we have seen, we suspect it is a terrific power that some day will be tapped to the benefit of humanity."[31]

But as inspiring as the notions of cosmic rays and atom building may have been to Millikan and the spiritually and scientifically like-minded members of the public, his scientific colleagues were considerably less impressed—when they bothered to take notice at all. "The atom-building hypothesis, which Millikan found so exhilarating, received remarkably little attention from his peers," Kargon remarked. "It generated at first little controversy, and even less support. For Millikan the cosmic photons 'must be in fact the birth cries of the elements,' but for others the entire effort seemed contrived."[32]

One reason was that, for the atom-building theory to be valid, cosmic rays had to be photonic in nature, rather than particles or corpuscles. But by the late 1920s, new instruments and techniques had arrived with which to study cosmic rays, so that physicists no longer had to content themselves with simply refining and improving upon the trusty electroscope and measuring ionization discharge.

The most exciting of these new tools were the Geiger-Müller (G-M) counter and the Wilson cloud chamber, both of which offered the possibility, at least after a fashion, to actually *see* cosmic rays, rather than just recording the indirect evidence of their presence.

Hans Geiger had first developed his counter about twenty years earlier, while a student of Rutherford. It was little more than a laboratory curiosity until Geiger and his own student Walther Müller produced a new and improved version in 1928. It consists basically of a metal tube or cylinder filled with an inert gas such as argon under low pressure, with a thin wire running down its length, insulated from the walls of the tube. The wire and tube are connected to a battery or other electrical power source so that they become two charged electrodes, one positive, one negative. When a charged particle such as an electron or proton passes through, it creates a path of ionized particles in the gas and thus a brief electric current that can be detected and displayed, whether as a blip on an oscilloscope or a click in an audio speaker.

That provided a useful way of detecting particles coming from, for example, a source of radioactivity such as a radium sample. German scientists Walter Bothe and Werner Kolhörster, however, saw another application: the study of cosmic rays. Their idea was to place two G-M counters, one above the other, with a layer of densely absorbing material between them, such as gold or lead. Any time both counters were triggered simultaneously, it would obviously indicate that a charged particle had passed through both. And by considering the miniscule time between detection with both counters and the known thickness of the intervening material, calculations of the particle's penetrating power, and hence its energy, could also be made. Of course, a certain number of random detections would be expected when separate particles happened to trigger both counters at the same time, but Bothe and Kolhörster were able to adequately filter out those occurrences. They called their technique the "coincidence method"—although in a sense what they were detecting were actually *anti*-coincidence events that were not the result of random chance but single, specific events.

Not long afterward, Bruno Rossi would come up with an ingenious modification of the Bothe-Kolhörster experiment that made it even more useful: he designed a "coincidence circuit" to connect two or more detectors together so that they would register a detection when triggered simultaneously. That innovation not only greatly enhanced the time resolution of the method, but also could be expanded to include a larger number of G-M counters in different configurations so that it would even be possible to figure out the direction from which a detected particle had arrived.[33]

Before the invention of Geiger-Müller counters and coincidence circuits, however, Scottish physicist Charles T. R. Wilson had invented a rather romantic way of studying radioactivity and detecting speeding particles: the cloud chamber. Enamored of the clouds and mists of his homeland, he became interested in studying the formation and behavior of clouds in the atmosphere. Since that clearly wasn't a subject that lent itself to controlled study, he set out to invent an apparatus he could use to produce clouds in the laboratory, a small, sealed chamber filled with a gas and water vapor, in which the precise control of temperature and pressure could create clouds through condensation.

That gave him his laboratory-based clouds, but he soon found another interesting phenomenon. Charged particles and x-rays from a radioactive source left trails of ionization in their wake, upon which water droplets would coalesce and leave a visible trail behind. You couldn't actually see the particle itself, of course, but you could see and follow the visible evidence of its passing, like a dust cloud left by a train in the desert or the wake of a distant ship seen from a seaside mountaintop. More than that, you could trace its path, see what happened, and what other particles might be thrown off when it struck something. You could place magnets outside the chamber and see how they would deflect charged particles and measure by how much, and thereby determine their energies. Best of all, you could photograph all of it to create a permanent record that could be studied at leisure.

Immediately after its perfection in 1911, the Wilson cloud chamber proved itself a revolutionary and versatile tool for physicists studying all forms of radioactivity. And in 1927, Soviet physicist Dimitri Skobeltzyn discovered yet another use for the cloud chamber: to study the behavior of gamma rays in a magnetic field, specifically the scattering of the electrons they produced through the newly discovered phenomenon of Compton scattering. "At that time, I was not interested in cosmic rays," Skobeltzyn recalled. "I was, however, aware of the work being done in the field."[34]

Two of his cloud chamber photographs showed the tracks of extremely high-energy particles, too powerful to be related to the gamma rays he was studying. He soon found likewise in other of his photographs. Eventually, the explanation became clear: for the first time, he had detected cosmic rays in a cloud chamber. Later, at a 1928 Cambridge University conference, Skobeltzyn said, "I demonstrated a collection of photographs of cosmic ray tracks and, I dare say, it produced some impression on the audience."[35] At the same conference, Bothe and Kolhörster's work was announced to be in progress, which no doubt stirred up more excitement, even if their results weren't quite available as yet.

By proving increasingly solid evidence that cosmic rays weren't photons but rather some type of particles, possibly electrons, this whole flurry of new experimental methods was beginning to pose some serious challenges to the whole idea of atom building and all the comforting philosophical and spiritual reassurance that it promised. After Victor Hess's exploits, matters of cosmic rays had remained a relatively quiet sideline of physics, while Einstein, Bohr, and the quantum physicists were making most of the noise and grabbing the headlines.

Now the new arsenal of experimental tools, along with the continued development of the tried-and-true method of high-altitude ballooning, was about to blow things wide open into not only a scientific controversy, but also a fundamental debate on the nature and fate of the universe that would take on much of the same tenor, passion, and apparent significance as a classic medieval religious war.

5

CHALLENGERS

"Fifteen prominent members of the California Institute of Technology have been murdered during recent months," announced a startling report in the May 23, 1930, issue of the *Los Angeles Times*. Among the dead were geologist John P. Buwalda, who had been found bound and gagged inside the aeronautics department wind tunnel, and internationally famous geneticist Thomas Hunt Morgan, locked in an ice box and frozen to death. "When we opened the ice box door . . . Morgan fell out with a clank, and little pieces chipped off and went skidding around the floor."[1]

At least Caltech head honcho Robert A. Millikan was not counted among the deceased. The writer of the murder accounts "believed that the famous physicist had made himself invulnerable by surrounding himself with cosmic rays."[2]

Fortunately for Caltech, however, there was actually no cause for alarm. The whole account was merely a joke story called "The Tech Murder Mysteries," published in *The Big T*, the annual student yearbook. Robert Millikan's personal reaction to the premature and exaggerated reports of his possible demise are not recorded, although the *Los Angeles Times* article mentioned that the yearbook editor "expressed appreciation for the faculty's breadth of vision which allowed publication of the annual without the slightest attempt at censorship in any form."[3]

As the new decade got under way, Millikan was relatively unconcerned with any attempts to murder him, whether professionally or in person. He was now over sixty, secure in his position as the grand eminence of American science and as the unquestioned head of America's newest center for scientific innovation. "He, who in his youth had rebelled against the dictatorship of professors and administrators, became by the irony of fate the most dictatorial of all dictators—though a benevolent one beloved of his subjects even when they chafed under his yoke," observed one of his students and later colleagues, Jesse W. M. DuMond, in a biographical essay.[4]

With the Great Depression sinking into the social and economic fabric of the nation, he also faced the daunting practical task of keeping Caltech going. His extensive network of professional colleagues, political chums, foundation presidents, and other influential types helped. Like nearly every other educational institution in the early 1930s, Caltech faced its rough economic patches, forced to postpone or cancel planned construction, new facilities, and new programs, and even faced bankruptcy at one point. But among his many virtues, Millikan was a supremely skilled administrator, adept at politicking and fund raising, and he managed to keep the Institute afloat.

He was also as yet not very concerned with the professional challenges gathering on the horizon from all the new activity with Geiger counters and cloud chambers. He felt assured in his scientific preeminence, dismissing most of the increasing questioning of his pet atom-building theories as inconsequential and misguided.

And the public was more than willing to accept his pronouncements as the ultimate authority on the subject, however they might be made. Millikan was set to deliver the final talk of Caltech's weekly public science series on June 6, 1930, on "Recent Cosmic Ray Experiments." While the lecture series was usually held in the assembly hall of the Norman Bridge Laboratory, this one was moved to the much larger Culbertson Hall nearby, since "the famous physicist is expected to attract a capacity audience."[5]

He did indeed, but the crowd of around four hundred people found upon arrival that "a talking-picture apparatus had been installed and their suspicions that the famous physicist had 'gone talkie' were verified when, after the grinding out of a musical overture, Dr. Millikan's likeness flashed on the screen and a voice, unmistakably his, began a serious discourse on cosmic rays," noted a report. "Even lectures on cosmic rays come 'in cans' these days," said the paper, explaining that his popularity had apparently made it necessary to agree to doing a six-reel "talking film" in which he announced again that "recent experiments have confirmed his theory that cosmic rays are the signals of the creation of common elements." Perhaps it was partly due to the fact that "talking film" was still quite a novelty at the time, and probably also because admission was free, but no one was reported to complain about the "canned" lecture.[6] No explanation for Millikan's failure to appear in the flesh was apparently offered.

No matter how much he continued to make such public pronouncements or how readily they would be accepted and celebrated in the press (and how often he would continue to be credited as "the discoverer of the cosmic ray"), Millikan would not, however, be able to ignore the work of his colleagues for much longer. The evidence was continuing to build, and it consisted not of theoretical speculations or arguments but of hard, physical, precisely measured data, just the sort of thing that Millikan had always valued so highly.

Some criticisms were easy to disregard, such as Sir Arthur Eddington's reported statements that cosmic rays "are the 'souls' of disintegrating atoms and in time the last erg of atomic energy will be gone. Unless we can turn time backward, [Eddington] says," referring to Millikan, "the planets and stars will gradually die down into cosmic dust."[7] Perhaps Eddington had been purposely trying to needle Millikan by using the same brand of spiritual imagery he liked to use in his talk of "birth cries," but Millikan was used to such jibes by now.

As 1930 progressed, however, it became increasingly clear that a new, more serious, and more determined challenger was in the wings, and this one would not be so easily dismissed. He was not only a fellow Nobel laureate, but unlike all those pesky Europeans, he was a fellow American, and almost as famous and prestigious as Millikan. The king of the cosmic ray mountain was about to face someone who, instead of simply nipping at his heels, might actually be able to knock him down off the summit. His name was Arthur Holly Compton.

For whatever reasons of chance and circumstance, Arthur Compton and Robert Millikan happened to have quite a lot in common. They were both midwesterners, although Compton was born not in Illinois but in Wooster, Ohio, in 1892. They were both the sons of Protestant ministers, although Compton's father was more than that, an accomplished academic as well who was a professor of philosophy and dean of Wooster College. His mother was a Mennonite devoted to public service and education. All of this endowed Compton with not only strong spiritual sensibilities that matched Millikan's, but also a solid love of learning and education and a belief in their importance.

But while Millikan's scientific inclinations didn't truly emerge until his undergraduate days, Compton started at least as early as age twelve when he became fascinated by astronomy, sitting on his front porch watching the stars come out. His father began to teach him the constellations and his older brother passed along an astronomy book from the family library. "I quickly sensed that I was on the track of something that I wanted to follow in greater detail," he recalled much later.[8] He ordered a telescope from Sears & Roebuck and soon became engrossed in observations, even photographing Halley's Comet when it reappeared in 1910. "I had at that time a strong emotional stirring that developed into an enthusiasm for scientific work," Compton remembered.[9] He began to think seriously about becoming an astronomer or some other type of scientist.

Compton's enthusiasm for astronomy was soon supplemented by a passion for airplanes and the new technology of aeronautics. "After making literally thousands of model airplanes, I next set out to plan a glider that could carry me," he wrote. He actually succeeded in constructing a twelve-foot long triplane with a twenty-seven-foot wingspan and eventually made several short flights in the craft in the summer of 1910. Around this time, his parents' gently enthusiastic support for their eighteen-year-old son's exotic and hazardous hobby sharply waned, as Compton noted. "Having demonstrated . . . my ability to get off the ground, Father and Mother urged me to stop further work on the project."

Compton himself had reached the conclusion that he had done about as much as possible with airplanes at that point anyway, so he was ready to move on. But it was already clear that he was definitely a hands-on type of guy, not content merely to sit on a porch swing and gaze dreamily at the sky in idle contemplation. If indeed he was to become a scientist, he was going to be the kind whose hands were dirtied not with ink or chalk dust, but with oil, chemicals, and sweat.

It was time to start thinking about college. For a short time, given his family background, he thought about the ministry, maybe even becoming a missionary. But although his parents liked such ideas, they sensed his true path lay elsewhere, noting his strong interest and abilities for science. "If I am not greatly mistaken, it is in science that you will find that you can do your best work," Compton recalled being told by his father. "Your work in this field may become a more valuable Christian service than if you were to enter the ministry or become a missionary."[10]

So the calling of Compton's life was set on science—but a science hand in hand with a religious sensibility, bringing both together not in opposition but in harmony. Although both men would eventually prove to have quite different conceptions of how that might be done, it was yet another common touchstone between Compton and Millikan.

Compton earned his undergraduate degree at Wooster College, then went on to graduate work at Princeton University, taught briefly at the University of Minnesota, worked for the Westinghouse company and the Army Signal Corps during the war, then won a fellowship to study abroad for a year at the University of Cambridge in Ernest Rutherford's Cavendish laboratory. It was here that he began the research in gamma and x-rays that would eventually earn him the Nobel Prize. Back in the States, he took up a professorship at Washington University in St. Louis and continued his experiments, culminating in a 1923 *Physical Review* paper that set out what became known as the "Compton effect" or "Compton scattering." Physicist Samuel K. Allison noted in a memoir that "Compton's discovery created a sensation among the physicists of that time."[11] Compton himself remembered, "When I presented my results at a meeting of the American Physical Society in April 1923, it initiated the most hotly contested scientific controversy that I have ever known."[12]

He didn't know at the time, of course, that he would spend much of the next decade dealing with an even more intense scientific controversy with the man whom he replaced when he moved from St. Louis to the University of Chicago later that year. Robert Millikan had decided to move to Pasadena permanently, and the university decided that Compton was just the man to fill the vacated faculty position. Compton continued to refine and improve upon his work on x-rays and the Compton effect, fending off scientific challengers who had tried to refute it.

In 1927 came his own Nobel Prize in physics. In the eyes of both the public and his fellow scientists, Compton had now become one of the very select group of America's supermen of science. That was an especially singular distinction in an age in which it seemed that America was still lagging behind its European counterparts intellectually and culturally, no matter its strengths in other realms such as industry and finance.

Like Millikan, Compton began to find himself treated more and more as a celebrity, sought after for learned speeches and profound statements about science and nearly everything else. He took it in stride, doing what he could when he could, but he was still essentially an experimental physicist at heart. Unlike Millikan, he wasn't the head of a major academic institution, someone who had to deal with raising money and making connections and building up the reputation of his school. Nor did he have Millikan's penchant for self-promotion or need for praise and professional affirmations. He just wanted to do his work.

And while Millikan's personal philosophies and spiritual inclinations had become fairly well set in stone by now, Compton remained open to further explorations in those areas. His 1926 sojourn in India would affect him profoundly, not only professionally through the work and contacts he would find with a new community of scientists, but also personally through a direct experience with a people and culture far different from those he had known in the United States and in Europe. "The experience of this year in India was to open my eyes," he recalled. "Years later I told my friends that it was the beginning of my education. It seemed to me that for the first time because of my foreign experience it was possible for me to see in new light the meaning of my own heritage."[13]

It's difficult to imagine the older, considerably more staid and traditionally minded Robert Millikan being similarly affected. "As a result of this trip my interests soon began to broaden into related sciences and the humanities," wrote Compton. "When I was a college student I had taken an active interest in philosophy, especially ontology, as taught by my father. During intervening years this interest had lain dormant."[14]

But no longer. The recent developments of quantum theory "seemed to have interesting philosophical implications. This, together with my broadening culture interests, made the philosophical implications of science occupy a considerable part of my attention in later years." Given Compton's background, that naturally also led to "a consideration of the relation of science to religion, a problem with which my father had wrestled, and which we had frequently discussed in my college days."[15]

The particular flavors of their philosophical musings and preoccupations are the place wherein the common religiosity of both Millikan and Compton begin to diverge, in ways that would ultimately and inevitably color their approach to the questions of cosmic rays and how they sought to answer them through their scientific work. "[Compton's] interpretation of the religious message of science . . . differed from Millikan's," wrote De Maria and Russo. "He [Compton] looked for the basis of human progress not in the cosmical evolution of the universe but rather in the freedom and rational capability of human beings." Also unlike Millikan, "Compton's optimism about science was not cosmic but rather practical and mundane."[16]

By the end of the 1920s, that practicality had brought Compton to a professional crossroads that seemed to be a natural progression from the pioneering work that he had done up to that point in his career. He wanted to get out of the laboratory and the classroom and continue to broaden his horizons. The recent work by colleagues such as Bothe, Kolhörster, and Rossi pointed to exciting new directions and what would have to be done next—by whomever was daring and ambitious enough.

Aside from the G-M counters and cloud chambers, one way to settle the question of whether cosmic rays were photons or particles would be if they exhibited a "latitude effect," a variation in intensity dependent on geographical location. Charged particles, positive or negative, would be deflected by the Earth's magnetic field in ways that could be measured. Photons, however, would not, so their intensities would have nothing to do with their distance from the magnetic poles.

During his 1920s peregrinations, before Millikan had finally decided for himself that cosmic rays had to be photons, he tried to look for some kind of latitude effect. But his efforts were somewhat half-hearted and marred by experimental problems. A Dutch scientist named Jacob Clay had in fact claimed to find a latitude effect in 1927, but his evidence was somewhat ambiguous and easy to disregard if one were so inclined, as Millikan certainly was.

If the existence or nonexistence of a latitude effect in cosmic rays was to be firmly established once and for all, somebody was going to have to take exhaustive measurements all over the globe, at all latitudes and at different altitudes and environmental conditions. That would mean more than a few trips here and there, a few balloon or airplane flights, or a long series of laboratory experiments. It would be a major scientific undertaking, unprecedented in its scope, number of participants, investment of time, and financial cost, requiring vast coordination of resources, organizational skill, and administrative acumen.

Now, in the new decade, despite a crippling economic depression, political upheavals, and widespread social malaise, the right person had come along to take it all on. Arthur Holly Compton, who had previously only dabbled somewhat in the cosmic ray arena and found it immensely interesting, had decided to get serious at last.

"Millikan loved publicity," observed *New York Times* science writer William Leonard Laurence in an interview. "In fact, somebody said once that a unit of publicity is a 'Kan,' of which 'Millikan' is a one thousandth part."[17]

There was no doubt that although he didn't necessarily spend much time chasing after it, Robert Millikan also didn't discourage the public accolades and press coverage that touted him as America's leading savant, not only on science but also on pretty much every other pressing issue of the day from unemployment to war and peace, the economy, and of course all matters religious. He was always prepared on every occasion to make just the kind of profound and authoritative pronouncements that journalists loved.

"One day I was just leaving Dr. Millikan's office in Pasadena after a long talk with him about the cosmic rays," newspaperman Bailey Millard noted in a radio address, "when I asked him if his discovery of these rays did not help to establish the theory of a cosmic or universal mind that ruled and directed all things. 'Why not say "God"?' he asked with a smile. 'There is no reason why a physicist should not believe in such a being or force. We use the word "God" to describe what is behind the mystery of existence . . . as providing a reason for existence and a motive for making the most of existence in that we may be part of the great plan of world progress.'"[18]

Millard concluded, "This scientist, like others, thinks that the cosmic rays result from the creative processes and are part of a great creative cycle. The theory that the process of degeneration and regeneration provides a never-ending cycle of creation gives us the hope that the material from which the universe is made as well as man himself has immortal life."[19] In a nation mired in the depths of the worst economic depression it had ever experienced, that was just the sort of thing that people wanted to hear.

And if one preferred to take their existential reassurances with more of a scientific flavor than a religious one, other journalists, particularly the science specialists such as Laurence, were happy to oblige. Laurence had just joined the *New York Times* in 1930, the new kid (although he was already past age forty) on the science beat, and he had something to prove.

So he went all out on one of his first big Sunday supplement assignments, in which he laid out the whole story on cosmic rays to date and the ultimate possibilities raised by Millikan's work and theories. "Out of the cosmic ray [Millikan] fashioned for science a new weapon with which to give battle to the dreadful Second Law of Thermodynamics, according to which the universe must come inevitably to destruction." Against the pessimistic, nihilistic doubters such as James Jeans and Arthur Eddington, wielding the "seemingly deadly weapon" of the Second Law, "Dr. Millikan goes forth in the shining armor of the cosmic ray," Laurence gushed. Making the by-then-common error of crediting Millikan with the discovery of cosmic rays, Laurence declared that it "may well prove to be the greatest scientific achievement of our age—a sign-post somehow pointing the way to the very doorstep of Creation's laboratory, in which matter itself is fashioned out of primeval electrons and protons."[20]

Laurence would discover that just as the greater public visibility and influence of scientists such as Millikan brought greater opportunities for science journalists such as himself to affect not only the public's view of scientists and their work, but also sometimes the work itself, that high profile also provided greater opportunities for getting into trouble. Several months later, he would inadvertently find himself playing the role of promoter and referee, if not the actual instigator, of a clash between two Nobel Prize winners.

As the ever-eager-for-good-news press trundled onward, and as Arthur Compton in Chicago continued to quietly put together plans for his own cosmic ray research, Millikan decided to go forth in the shining armor of the cosmic ray to the Arctic, along with Caltech secretary Edward C. Barrett. "Although the trip was characterized as a vacation for both men, it was learned that Dr. Millikan's luggage included five large packing cases filled with intricate scientific instruments,"[21] reported the *Los Angeles Times* on August 16, 1930. Millikan was planning more high altitude—and high latitude—electroscope measurements.

The 5,000-mile journey would take the men all the way up to fewer than 500 miles from the Arctic Circle on the western shores of Hudson Bay, where a construction camp called Fort Churchill served as home base. "The Pasadena men report that they were given very comfortable accommodations in a bunkhouse reserved for officials," noted the press. Aside from Fort Churchill, Millikan and Barrett took readings at several points along the way in Winnipeg and elsewhere. "I was very pleased at conditions for observing at Churchill but details of the results have not been worked out," Millikan reported.[22]

Millikan returned from the far north to Pasadena just in time for the opening session of the National Academy of Sciences meetings, at Caltech's Dabney Hall on September 22. Millikan spoke first, describing results from his recent northern expedition and coming up with another cosmic ray theory that, not coincidentally, would help to explain away talk of a latitude effect and that cosmic rays had to be charged particles. This new theory introduced "earth spots," which he explained were "slits or holes in the earth's atmospheric envelope. They result when the air, expanded by the sun's warmth, rises and boils over at the upper limits of the earth's air blanket."

This would supposedly result in a diurnal variation in the cosmic ray flux, with more rays reaching the surface in the warmer afternoons than the cooler nighttime "because the air blanket has been partially moved locally by this heated spot," Millikan held. And since barometric pressure was also affected by air temperature, that meant that a cosmic ray electroscope could be used as a weather instrument.

"[It] is a simpler and a more fundamental instrument than the barometer and I expect it to be an aid in bringing about advances in weather forecasting and ultimately to find a place in meteorological stations," Millikan told his audience. "The air is simply an absorbing blanket interposed between us and a constant source of radiation coming into the earth uniformly from all directions, and every eruption, or wave, or ripple in that blanket is accurately reflected by this cosmic ray electroscope."[23] It may have been an interesting if somewhat fanciful idea, but the fact that electroscope readings are not part of nightly weather forecasts in the twenty-first century demonstrates the fate of this particular Millikan inspiration.

One perquisite, or perhaps occasional disadvantage, to being a world-famous scientist also in charge of a major local educational institution was that an enthusiastic local press might follow one's personal ups and downs a bit too closely. The *Los Angeles Times*, always eager to report on any doings of the hometown science hero, made it a point to dutifully cover Millikan's brief November bout with bronchitis, noting that while he had been confined to bed, his condition was not serious according to his doctor. "This summer, Dr. Millikan's strength was taxed as the result of his cosmic-ray studies on the shores of Hudson Bay, Can.," the paper explained helpfully.[24]

He recovered quickly enough to explain to a New York meeting of life insurance executives that the crippling unemployment then besetting the United States couldn't be blamed on science. "The task of science is to supply as many legitimate human wants as possible with one foot pound of energy, and to extract the maximum of satisfaction to the race out of our present reserves of energy," Millikan told the insurance salesmen. In a bit of scientific prescience, he observed that in the future, "science will be in a position to supply our wants by using direct sunlight instead of stored solar energy in the form of oil and coal." He also managed to work some rather oblique references to cosmic rays into his talk, noting that they couldn't provide a power source for humanity as some had speculated.[25]

By the end of December, one of the few scientists more famous and celebrated than Robert Millikan was also in New York. Albert Einstein had returned for his second visit to America, following his first in 1921. After touring the sights of the Big Apple, he would head to California on New Year's Day for his first venture west of the Mississippi. His ship would dock in San Diego, where he would be met by Arthur H. Fleming, Caltech's president of the board of trustees, and driven up to Pasadena, where he and his wife Elsa would stay in a rented house whose location was a carefully guarded secret. Among many other planned activities, including lectures, tours, and visits, Einstein would "compare notes . . . with Dr. Robert A. Millikan on possible important theoretical deductions from Dr. Millikan's cosmic ray research."[26]

For Albert Einstein, the entire experience would be just another chapter in his long and storied career. For Robert Millikan, it offered further opportunities to promote and validate not only Caltech, but also his own ideas about the universe. Or so he hoped.

6

MOUNTAINEERS, FLIERS, AND SAVANTS

Albert Einstein was aboard the *S.S. Belgenland* somewhere in the South Pacific en route to San Diego, dictating notes and lounging on the veranda of the ship's cafe in pajamas.[1]

Robert Millikan was several thousand miles away in Cleveland, Ohio, at the annual meeting of the American Association for the Advancement of Science (AAAS), where, as the retiring president, he would deliver the opening address. It provided yet another opportunity to strike a blow against the proponents of the Second Law of Thermodynamics and "heat death," and to argue in favor of atom building and atomic birth cries.

It is doubtful that the content and tone of Millikan's address came as a surprise to most of those present. He systematically laid out "the history of the development of scientific evidence bearing on the question of the origin and destiny of the physical elements," describing ten significant discoveries and developments of the past century with some bearing on the problem. These included the conservation of energy principle, the Second Law of Thermodynamics, evolution, and radioactivity, which "raised insistently the question as to whether the creation, or at least the formation, of all the elements out of something else may not be a continuous process—a stupendous change in viewpoint the discovery of radioactivity brought about, and a wholesome lesson of modesty it taught to the physicist."[2]

Some in the audience might have chuckled inwardly at hearing Robert Millikan, of all people, speaking of modesty. He continued, inevitably bringing his presentation around to the atom-building theory, elemental birth cries, and the evidence of all contained within cosmic rays, also describing his recent work on the subject.

"This has been speculatively suggested many times before in order to allow the Creator to be continuously on His job," Millikan intoned. "Here is perhaps a little bit of experimental finger-points in that direction."[3]

In what may have actually been some degree of modesty but more likely was simply a bit of scientific circumspection, Millikan went on, "But it is not at all proved nor even perhaps necessarily suggested. If Sir James Jeans prefers to hold one view and I another on this question, no one can say us nay. The one thing of which you may be quite sure is that neither of us knows anything about it."[4]

He concluded, "I am not unaware of the difficulties of finding an altogether satisfactory kinetic picture of how these events take place; but acceptable and demonstrable facts do not, in this twentieth century, seem to be disposed to wait on suitable mechanical pictures. Indeed, has not modern physics thrown the purely mechanistic view of the universe root and branch out of its house?"[5]

The Associated Press reported that Millikan also noted that a belief in evolution didn't necessarily imply disbelief in religion. "Neither evolution nor evolutionists have in general been atheistic—Darwin least of all." To the contrary, evolution served to "identify the Creator with the universe."[6]

Among the audience, along with many of his journalistic colleagues, was *New York Times* science writer William Laurence. Millikan had certainly said little that Laurence and his fellow science reporters hadn't already heard before, far more than once. Still, even if it was lacking any startling new revelations or discoveries, the opening address was an important part of the proceedings. So following standard procedure, Laurence dutifully cabled a brief summary back to New York before writing up the full story later, which included this account: "Professor Millikan, Nobel Prize winner, in his presidential address tonight before the largest scientific society in America, said that the cosmic rays are the birth cries of atoms being born in interstellar space and that they present the first experimental evidence that the creator is still on the job."[7] To Laurence, a confirmed agnostic if not outright atheist, religious skeptic, and hard-bitten newspaperman, it was routine stuff, not necessarily very convincing, and worth mentioning only out of journalistic completeness.

The reaction was far less blasé among the night staff back home on West 43rd Street, however. "The night managing editor of the *New York Times* at the time was a devout Catholic, and his three assistants—two of them were born Catholics, and the other was a converted Catholic," Laurence recalled. "Well, when that hit the desk I can see those Catholics . . . it certainly made a sensation." It seemed to them that Laurence was filing a story about the first firm scientific proof of the existence of God.

The night editor immediately cabled back asking for at least a two-thousand-word story, then doubled that to four thousand. Before Laurence had penned the first thousand words, another cable arrived demanding the full text of Millikan's speech.

Full text? Laurence wondered. What the hell was going on? The text of a speech might be printed in full if it was from the President of the United States—maybe—but not from some scientist, even a famous Nobel laureate. But it wasn't up to him, so Laurence stayed up late, wrote his four thousand words, sent the story along with the full Millikan address off to New York, and finally went to bed.

Next morning, as Laurence settled down to breakfast with the morning edition of the *New York Times*, he nearly choked on his coffee.

There it was on the front page: "Millikan Finds Creation Still Goes on While Creator Directs the Universe," declared the headline of Laurence's story. It wasn't the lead headline, but it was close, and the entire story filled six columns including the text of Millikan's speech. As if the headline wasn't bad enough, there was the subheading: "Belief in Evolution Does Not Conflict with Religion, [Millikan] Tells Scientists at Cleveland." The story continued on page 12 with the heading "Millikan Believes World Has Creator."[8]

After barely a year at the *New York Times*, William Laurence had broken the story of the century: Science Finds God. He realized he had no one but himself to blame. He had put Millikan's line about "the Creator still on the job" front and center, both in his initial summary and in the first paragraph of his final story, because Millikan <u>had</u> said it, and it was Laurence's job to report it.

As had happened with similar Millikan pronouncements in the past, religiously-inclined editorial boards across the country feasted on these latest proclamations. "Perhaps he really announced little additional to previous reports on his discoveries," the editors of the *Philadelphia Inquirer* accurately observed. "But his picturesque phrase has turned fresh attention to his adventures in the realm of physics. . . . Thousands of years ago the Psalmist felt assured of a certain Presence whether he ascended into heaven, made his bed in hell or took the wings of the morning to dwell in the uttermost depths of the sea. With somewhat more precision modern science has attempted to explore the same regions, and only in the icy depths of a Rocky Mountain tarn could Millikan testify that the cosmic ray did not function. The layman need scarcely inquire why. The detail has not bothered the physicist in recognizing the creative force of the universe." Millikan, in other and far fewer words and without really saying it, had indeed found God.[9]

For Laurence, it was an important lesson about the power of the press, the power of a managing editor, and most importantly his own power and responsibility to make certain that what he wrote was what he meant. It was not enough to scrupulously record facts, ask good questions, and get complete answers. In choosing what to emphasize and what to downplay, he was affecting how a story might be received and interpreted.

Laurence quickly recovered from any embarrassment he may have felt. "That established me very well with Millikan; it established me with the *Times*; that really was a big story," he said.[10]

Two years later, during his next significant journalistic encounter with Millikan, he would have cause to regret such a sunny assessment.

Albert Einstein found his arrival in Southern California to be a welcome anomaly. For once, instead of being met by brass bands, adoring throngs, and shouted questions from reporters, he was left alone.

Mostly, at least. He did take a few questions from the press because he'd realized by now that such was unavoidable. One of them would have pleased the absent Millikan: "Do you agree with Dr. Millikan that belief in evolution does not necessarily imply disbelief in religion?" Well accustomed to such queries by now, Einstein's response was polite and measured: "There is no conflict. It all depends on what constitutes your view of religion—your faith."[11]

His hosts at Caltech reportedly were "exerting every effort for Dr. Einstein to 'be himself' during his visit." Aside from several special invitation-only affairs, there would be no public receptions, dinners, or other events. Instead, when not being left alone to work in his rental house, the great man would be quietly ushered to tours and visits with scientific colleagues, including Albert Michelson and Robert Millikan.[12]

Still en route home from his trip to Ohio for the AAAS convention, Robert Millikan wasn't present yet, no doubt to his enormous frustration. This first Einstein visit to California, as well as those that would follow in the next two years, "capped Millikan's campaign to make Caltech one of the physics capitals of the world," wrote Caltech archivist Judith Goodstein.[13] Einstein and Millikan finally got together on January 6, 1931, as noted in Einstein's travel diary: "It is very interesting here. Last night with Millikan, who plays here the role of God."[14] The scientific genius was also extremely perceptive.

The next day, Millikan along with his old University of Chicago boss and mentor Albert A. Michelson officially welcomed Einstein to Southern California in a simple ceremony. "We have here working together in friendly co-operation three institutions of research—the California Institute of Technology, Mount Wilson Observatory and the Huntington Library and Art Gallery," Millikan said to Einstein. "We wish you to feel free to make full and complete use of the research facilities offered by any and all of them." Einstein graciously acknowledged the invitation, after which "the group of savants then stepped to the veranda and chatted while posing for newsreel photographers," reported the newspapers.[15]

Einstein remained in California until about the end of February, working when possible, meeting with colleagues publicly and privately, catching and recovering from a slight cold, and being shadowed by the press despite the best efforts of his hosts to protect their guest from undue attention. Practically every move, every word, every gesture from Einstein and his wife was duly noted, recorded, reported, remarked upon, and analyzed for cosmic significance, whether mundane or scientifically related.

For the California Institute of Technology, Einstein's California sojourn certainly did much to cement its reputation and international prestige. A singular genius such as Albert Einstein didn't waste so much of his precious time hanging about any place that wasn't more than worthwhile.

For Robert Millikan, Einstein's visit was perhaps somewhat less directly advantageous. Although any sort of direct exploitation of Einstein for personal advantage was farthest from Millikan's mind, he nevertheless managed to work in some connections between Einstein's achievements and his own ideas. In a tribute dinner at Pasadena's Athenaeum in the middle of January attended by a slew of distinguished guests and scientists, including Michelson, astronomer Edwin Hubble, and physicist Richard Tolman, toastmaster Millikan found a way to insert himself, building on Einstein's own remarks.

Einstein spoke briefly to the group, noting the connections between the work of the scientists in attendance and his own, including Millikan's. "I acknowledge gratefully Dr. Millikan's researches concerning the photoelectric effect," he said, and its implications "for the corpuscular structure of radiation." In his own comments, Millikan noted that Einstein's photoelectric equation "necessitated our return to at least a semicorpuscular theory in the nature of radiant energy," and that it along with several other aspects of Einstein's theories constitute "the most important basis for the cosmic ray conclusions that I am now wishing to draw."

Whether Einstein and the other scientists present agreed with that or not was another question. But in any case, even if Einstein was most celebrated as the originator of the theory of relativity, Millikan remarked that one could throw relativity into the waste basket and "Professor Einstein's position as the leading mind in the development of our modern physics would still remain unchallenged."[16]

Probably so. But Millikan's position in modern physics remained quite open to challenge.

One challenger arrived in Pasadena at the end of April 1931, while Millikan was away again on another trip back East. This was the distinguished British astronomer and physicist Sir James Jeans, with whom Millikan had already clashed, more or less politely, in the press and scientific literature.

Jeans took considerable exception to Millikan's pet "birth cries" ideas, holding that cosmic rays signified not creation but annihilation. But he was less personally invested in his judgments on the question than Millikan, who liked to imbue them with deep philosophical significance. For Jeans, the issue was one of science, not spirituality. "It may be that both of us are wrong but I am looking forward to discussing the question with Dr. Millikan," Jeans told reporters in his Oxford accent. "Dr. Millikan and I probably are in agreement more than 95 per cent of the time and while we may have some warm discussions regarding our 5 per cent difference, they will be in the friendliest vein."

He laughed off a London newspaper labeling him a "pessimist" because of his talk of the universe running down. "Science isn't much interested in whether its findings are optimistic or pessimistic—it's the truth we are after. Even if it is true that the universe is getting older, I don't think it's anything to worry about or be gloomy about."[17]

While the press naturally liked to focus on arguments between world famous scientists, Jeans' discussions with Millikan would only be incidental to the actual purpose of his visit, which was to meet and consult with his colleagues at the Mount Wilson Observatory. Jeans and Millikan finally met up on May 6 at the observatory, shook hands and posed for some pictures. Millikan stood dark-suited in a bow tie, hands folded at his waist, while Jeans stood hands on hips in a lighter gray ensemble, looking at the camera with a rather challenging expression compared to Millikan's neutral facade. Neither man was smiling.[18]

Unlike Einstein's more leisurely stay in Pasadena, Jeans's visit would last only a few days, because he was due to receive a medal from the Franklin Institute in Philadelphia before returning home to England. Still, there would be time for a public discussion (or debate, if one preferred the term) between Millikan and Jeans on their respective theories. Or as the Associated Press phrased it, "two mighty huntsmen of the universe, whose scientific sport has been tracking the cosmic ray, [coming] together . . . to swap tales of their pursuit."[19]

Jeans went first, speaking on "The Origin of Cosmic Radiation," laying out in detail the latest discoveries of Edwin Hubble and Milton Humason demonstrating the existence of the red shift and an expanding universe. "The results of all this are absolutely sensational; they show that practically all the nebulae are stampeding away from us at terrific rates," he told the audience. Jeans was in fact setting out what would later be recognized as the first definitive evidence of the Big Bang and the origin of the universe.

And if the universe had a definite beginning point, that implied that it would also eventually have an end. Jeans explained, "Not only is the amount of matter in the universe decreasing, but what is left continually spreads itself further and further apart . . . the material universe appears to be passing away like a tale that is told, dissolving into nothingness like a vision."[20]

Those, of course, were fighting words to Millikan, who proceeded to offer his by-now-familiar pitch about birth cries of atoms in cosmic rays. It was left to the audience to determine who was the optimist and who was the pessimist, but clearly neither Millikan nor Jeans were persuaded to change their minds. As the Associated Press observed, Jeans was "ready to start his homeward journey . . . more convinced than ever that the universe is going to pieces. He finds nothing illuminating in Dr. Robert A. Millikan's belief that the cosmic rays tell a story of the building of a bigger and better universe."[21]

Quite apart from any thoughts of cosmic creation and annihilation, Jeans's visit ended on a very human note of darkness and mortality: Albert Michelson, first American Nobelist scientist, colleague to both Jeans and Michelson, and mentor, advisor, and role model to Millikan, passed away at his Pasadena home on May 9 at age seventy-eight. He was a monumental figure to the public and to scientists around the world, but especially to Millikan, who told the press, "I personally owe everything to the fact that thirty-four years ago, Dr. Michelson took me as a new egg into the nest over which he brooded at the University of Chicago."[22]

Millikan and Jeans would pick up where they had left off in September, when Millikan stopped off in London on his way to the Rome nuclear physics conference. This time Jeans not only had home field advantage but was joined by a powerful ally in British astronomer Sir Arthur Eddington, who had also made significant contributions to the expanding universe concept. This time, before a much larger crowd of about two thousand, the discussion reportedly became somewhat more heated than the Pasadena meeting.

Against the august scientific status of Jeans and Eddington, Millikan also had in his corner some heavyweight backing, if not of an altogether scientific variety, including a British general and a bishop of Birmingham, who shored up Millikan's technical arguments with philosophical cant and reassurance. "Their criticism so stung Jeans that at the end of the meeting while a crowd of 2000 was streaming from the hall he asked to be allowed to defend himself," noted the *New York Times*.[23]

Predictably, Millikan's hometown paper awarded him the engagement. "From the report of the discussion it is plain that the Pasadena physicist had the better of Jeans and Eddington and the audience evidently thought so, for it hardly waited to hear the concluding argument of Jeans, though, of course, laymen might hardly be expected to cast a deciding vote."[24]

Although Millikan didn't realize it at the time, the encounters with Jeans were all merely warm-up bouts for the main event, which would come just over a year later, in 1932.

French physicist Pierre Auger, who would later make his own significant contributions to cosmic ray science, noted in a popular text, "In the course of its history every science passes through one or more 'heroic' epochs." By 1931 the "heroic epoch" of the cosmic ray field was coming into full bloom, borne on the shoulders of, as Auger called them, "mountaineers, miners, divers, and fliers."[25]

On the morning of May 27, a member of the flying contingent would sally forth. Swiss-born physicist and balloonist Auguste Piccard took his enclosed aluminum gondola to about 50,000 feet or just under ten miles, setting a new world altitude record. From a scientific standpoint, it was hardly a significant achievement, as the flight was so plagued with problems, including an oxygen leak and overheating of the gondola, that "scientific experiments were out of the question," as historian David DeVorkin noted.[26]

Robert Millikan thought the Piccard flight was "a most interesting venture" that he hoped would lead to further similar stratospheric observations but noted that he had already made extensive cosmic ray measurements at similar altitudes with unmanned balloons. Maybe it wasn't all a waste of time, however. Some of Millikan's colleagues also speculated

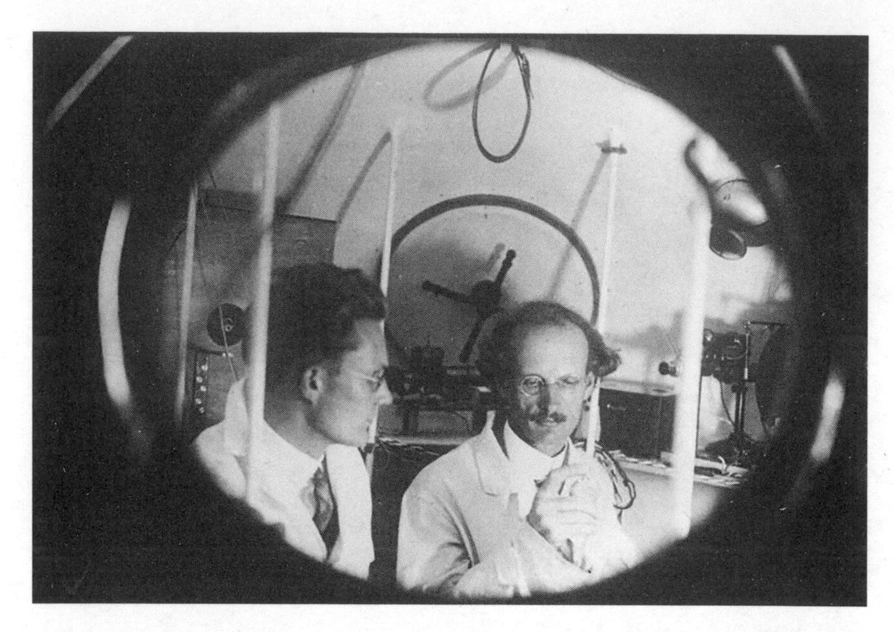

6.1 Auguste Piccard (right) and Paul Kipfer inside their gondola, 1931. (Wikimedia Commons, German Federal Archive)

that Piccard's data might end up providing support for Millikan's recent ideas about earth spots and cosmic ray effects on meteorology.[27]

While Piccard was setting records, Arthur Compton personally joined the ranks of adventurers as one of Auger's "mountaineers." By the previous fall, he had gathered the scientific colleagues who would form the core of his dedicated cosmic ray program, including his former Ph.D. student and now MIT professor R. D. Bennett and University of Denver professor J. C. Stearns. They had spent months designing and building a precision ionization chamber that could operate over a wide range of temperatures and pressures, just the sort of thing needed for a campaign of measurements in all sorts of geographical locations from mountaintops to underwater. In the late summer and early fall of 1931, they started to take it out into the field for some serious test drives.

After some preliminary measurements around the University of Chicago campus in south Chicago, Compton, Bennett, and Stearns loaded up a ton and a half of equipment, much of it consisting of an 1,800-pound lead shield for their electroscope, and drove to Mount Evans, Colorado. They spent several weeks atop the 13,000-foot peak collecting readings, looking for any variations arising from temperature and atmospheric pressure as well as time of day, which was a controversial question in the cosmic ray community.

In a letter to *Physical Review*, Compton, Bennett, and Stearns noted that some previous experiments, including those of Millikan, had appeared to show diurnal variations in cosmic ray intensities: "This is the type of apparent variation that one should expect if the apparatus is not kept at a uniform temperature." Other researchers believed such variation existed only in the "softer," that is, less energetic and less penetrating cosmic rays. "These softer components, however, can only be studied in very high altitudes where the temperature variation between day and night becomes relatively large," wrote Compton and his colleagues. "We, therefore, determined to study the variations in the intensity of cosmic rays in a high altitude in such a way that possible temperature variations would not influence our results." They found no intensity variations within the margins of statistical error, concluding that "it would appear that probably some of the previous changes that have been recorded may be due merely to variations in the temperature of the apparatus employed."[28] Or in other words, somewhat sloppy experimental technique.

The Denver press took note of the expedition both because Stearns was a local notable and because scientists lugging exotic equipment up mountains wasn't an everyday occurrence. Not all of the coverage was serious, though. The *Denver Post* published a photo of a group of "pretty Denver university coeds" gathered around an electroscope with Stearns, supposedly inquiring, "Oh, Professor, tell us, are there any cosmetics in cosmic rays?"[29]

Afterward, Compton spent only a brief time back in Chicago before setting off again, this time farther afield to the Jungfrau in the Swiss Alps, where he would take more electroscope readings to supplement the Colorado work, assisted by University of Zurich physicist Marcel Schein. Again, allowing for some differences in local conditions, Compton found no significant variations with altitude.

Then it was off to Rome and the nuclear physics conference, with Mussolini's unwelcome presence, and Compton's minor epiphany at Bruno Rossi's cosmic ray presentation. Here was confirmation that he was indeed on the right track. The golden road to discovering "the secret of the cosmic rays," as the newspapers kept calling it, lay in the continuation and expansion of what he had already started: an exhaustive worldwide survey of observations across all the continents and in every environment.

That "secret" had already gone through a number of permutations by the early 1930s, beginning with the subtle manifestation of a mysterious ionization or "penetrating radiation" from everywhere and nowhere, then the realization that some sort of new unknown radiation was responsible. Victor Hess, his European colleagues, and Robert Millikan had finally established its reality and identified its general cosmic origins. Now the remaining "secrets of the cosmic ray" were not whether it existed or where it came from, but what it actually was and what created it.

Those were the secrets that Arthur Compton would probe through the most ambitious worldwide scientific collaborative effort to date.

Assuming, of course, that he could get the money to pay for it.

Anyone who followed scientific developments in the early 1930s, whether avidly or casually, likely thought of Robert Millikan as Mr. Cosmic Ray. Whenever talk of cosmic rays came up in the news, in editorial musings, radio programs, newspaper advertisements, or even comic books, Millikan's name would be the one most often invoked.

That was the public perception, but scientifically as well, at least in America if not in Europe, Millikan was considered by most to be the reigning monarch of cosmic rays, the leading expert, the final authority. One might not agree with his every pronouncement and opinion, but in any case, his preeminence in the field could not be denied.

That dominance extended into the crucial domain of funding. Before World War II and the postwar advent of Big Science funded by governmental agencies, much if not most significant scientific research was privately funded through foundations and other similar bodies set up by wealthy tycoons and industrialists such as John Rockefeller and Andrew Carnegie. It was through the beneficence of the latter's largesse that Millikan had supported almost all of his cosmic ray research, particularly the various travels.

Not surprisingly given Millikan's reputation and his long-established relationship with it, the Carnegie organization also tended to automatically equate cosmic rays with Millikan, to the point where anyone else requesting support for such research was going to face inevitable comparisons and concerns about duplicated effort. This was precisely the problem that Arthur Compton found himself facing when he approached the Carnegie Corporation in 1931.

It was, of course, the Depression, and science and scientists were just as hard hit by the economic realities of the time as everyone else, at least professionally. Over a quarter of the expenses for the Colorado and Swiss Alps expeditions had come out of his own pocket. "I remember not getting a winter coat because we put our own money into cosmic rays," remembered his wife Betty, noting that Compton's employers at the University of Chicago "didn't think cosmic rays were anything that was very vital." Betty harbored no resentments, however, saying that "the brightest idea Arthur ever had was to get himself out of the basement laboratory room watching a spot of light proceed across a scale and realize that the earth was a big magnet and he could take readings in different parts of that big magnet and get his information, and also see the world."[30]

But getting out of the lab basement and seeing the world would take money. Compton first approached some smaller foundations as well as a Chicago bank president who had been considering establishing some sort of scientific foundation, but with no success. For a little while, it seemed that Betty Compton was going to continue to go without a new winter coat again.

Arthur Compton had better luck with the Carnegie Corporation when he wrote to its president in March 1931, requesting $18,000 for "a study of cosmic rays at high altitudes and at various magnetic latitudes."[31] Most of the money would be used for travel expenses to South America and Alaska along with supporting equipment.

Although it was a more than reasonable request, it placed the Carnegie people in something of a quandary. They had already been funding Millikan's cosmic ray work for about a decade, and he was the prominent American researcher in the field. Now here was a relative newcomer to the cosmic ray game, Nobelist or not, requesting Carnegie dollars. What if Compton's work was basically going to end up duplicating the work the Millikan was already doing and for which Carnegie was already paying?

An internal debate, or at least discussion, ensued among the Carnegie Institution leadership. Compton was invited to meet with Carnegie officials to discuss his plans and whether they might complement or duplicate Millikan's well-established efforts. According to historians De Maria and Russo, Compton found a strong ally in J. A. Fleming, the director of Carnegie's Department of Terrestrial Magnetism. That department had done some previous cosmic ray work of its own and was very interested in exploring ways to combine laboratory work with field observations, in just the way that Compton's proposed program would undertake. After meeting with Compton, Fleming advised the Carnegie president to fully support the work, which "would be much more than a repetition and confirmation of the work done by Millikan and would approach the whole subject from a wider point of view and in a thoroughly systematic way."[32]

The Carnegie people chewed on the question for another six months before finally officially informing Compton at the end of November that he was getting his money. As Carnegie president J. C. Merriam told his board of trustees, "At first it was thought that perhaps Dr. Millikan had a monopoly of the study of cosmic rays, and it occurred to us after thinking about the matter that he ought not to have a monopoly."[33]

Robert Millikan may have had some other ideas, of course, but it didn't matter. Arthur Compton had the money, the resources, and the people he needed, and was now in a position to get down to some serious planning.

He was not about to waste any time.

7

TRAVEL ARRANGEMENTS

At the end of 1930 and the beginning of 1931, the newspapers were filled with headlines about Robert Millikan (if not Albert Einstein, or often, both together): "Evolution Theory Is Not Atheistic, Declares Millikan"; "Millikan Finds Creation Still Goes On"; "Dr. Millikan Addresses Noted Scientists," and so on. Particularly in the Los Angeles newspapers, in which coverage of anything Caltech-related would also invariably involve a mention of "the Chief," if just in passing.

As 1932 dawned, it was Arthur Compton's turn in the press spotlight. Not that he was a stranger to the newspapers, of course; he was, after all, a Nobel Prize winner, which made at least his professional doings noteworthy by definition. But now he was preparing to do something wholly unprecedented, launching an undertaking that was not only enormously interesting from a scientific perspective, but also irresistibly romantic and adventurous even to the types who never read anything except the sports pages and stock market returns.

"A world-wide expedition bent on discovering the baffling secret of the cosmic ray will be undertaken this year by Dr. Arthur H. Compton, Professor of Physics at the University of Chicago, Nobel Prize winner in 1927 and distinguished for his research in the field," proclaimed the *New York Times* smack in the middle of its front page on Sunday, January 3, 1932. The article, as well as numerous others appearing nationwide and internationally

7.1 Robert Millikan (left) and Albert Einstein, 1932. Einstein made several visits to Caltech during the early 1930s. (Wikimedia Commons)

around the same time, spelled out the broad outlines of Compton's plans and travels, with ample commentary from the man himself.

"A survey such as this should give the most adequate test that has yet been devised to distinguish whether the cosmic rays are photons as are light and X-rays or electrons such as give rise to the earth's aurora," the piece quoted Compton. Naturally, the work of Millikan was mentioned, especially the fact that Compton's recent mountaintop measurements seemed to "agree in general" with Millikan's own experiments. There was also the obligatory nod to Millikan's atom-building ideas, though Compton was noncommittal.

No article about cosmic rays would be complete without a bit of philosophical speculation, and the *New York Times* didn't disappoint. Compton's musings, however, tended to hew closer to science than to religious woolgathering. Speaking of the implications of quantum physics and Werner Heisenberg's recently formulated uncertainty theory, Compton observed that "the uncertainty relation is a thing that represents real limitations of our physical knowledge. . . . The laws of the new physics cannot predict an event; they tell only the chance of its occurrence. As one whose experiments are partly responsible for this dramatic reversal of the physicists' point of view, I have been especially interested in tracing what the significance of this change may be to human life and thought."[1]

He wasn't the only one. Writing in the *Chicago Tribune*, Philip Kinsley enthused, "Science has a New Year's message of good news to present," one that would give "a glimpse of a new golden age of humanity, a future in which man may become the master of his destiny, instead of the victim of an unreadable, whimsical fate." Kinsley's article quoted Compton's scientific and some of his philosophical musings at considerable length, blending them together to show how the work may provide a "cosmic clew to human destiny," as the headline states. "It is no longer necessary from a scientific standpoint, Prof. Compton believes, to consider this universe as a place of chaos and night, with mankind sailing aimlessly along desolate shores and perilous seas, with certain doom ahead," Kinsley wrote. "It is, on the contrary, permissible, on this same basis of science, to postulate a fundamental unity and order, and to think of all living and perhaps nonliving things as well as controlled by something approaching consciousness, something greater than the individual." Kinsley acknowledged that such speculations were ranging far afield of science into the realm of religion. "But Prof. Compton does not flinch at that. He would, however, substitute understanding for faith."[2]

Meanwhile, the editorial page of the *Boston Globe*, in an example of somewhat tormented literary fabulism, seemed to question the ambitions of scientific exploration and the motives of those who provide it with material support. Science, said the *Globe*,

> begins to exact the superstitious veneration formerly accorded to more ancient deities. When a scientist proposes a series of expeditions to the top of high mountains in Peru, New Zealand, Australia, Hawaii, Alaska, and Switzerland in the hope

> of ascertaining something definite about the "cosmic ray," no man who does not pretend to expert knowledge of cosmic rays can say him nay should some rich gentleman poke up the money for the trips. . . . If I say I have some mysterious electrons up my left cuff which I alone understand, whereas you have lots of money, I may inform you that given some of your money I might (just possibly might) be able to release atomic energy or something like that, which would make you all-rich or all-powerful, and wouldn't that be wonderful? Aren't you willing to gamble a few millions?[3]

It's unclear whether the editors are criticizing scientific hubris, venality, or both. "Probably there is but little doubt that these gentlemen will come back with a first-class cosmic ray if given their expense money for the six-fold mountain climbing expedition. And it makes a nice ride for them."[4] The facts that Compton had already secured the funding for his project, that it hardly amounted to "a few millions," and that in any case it had come from private sources rather than the pockets of public taxpayers, seemed to have escaped the editors.

Whether or not one was inclined to use the subject of cosmic rays as a springboard for fanciful flights of religious and philosophical speculations, the whole story of Compton's planned campaign had a near-universal appeal. To most people in the early 1930s, scientists were indeed, as the headlines often called them, "wizards" or "savants" or "geniuses," working unimaginable experiments while secreted in their laboratories filled with exotic equipment and apparatus, doing incomprehensible things and exploring questions that mere mortals couldn't hope to understand, communicating with each other through the abstruse language of mathematics, and only occasionally deigning to emerge from their labs or down from their mountaintops and ivory towers to share the mysteries of the universe with the rest of us. They were serious, sober, well dressed (save for Einstein with his old sweaters and wild hair, but he was allowed to be the exception because he was a genius), and they spoke in measured, learned tones about measured, learned subjects. If one wanted an exemplar of the type, one had to look no farther than the grey-haired, straitlaced, patrician Robert Andrews Millikan.

But Arthur Compton didn't fit the mold. Whereas Millikan was an old, uptight guy well into his sixties who hobnobbed with presidents, prime ministers, and industrial bigwigs, Compton was a young, vigorous, handsome guy still in his thirties with an attractive, vivacious wife and a young son who accompanied him on his expeditions, now preparing to travel across the globe, not to big cities and comfortable salons but to remote mountaintops, jungles, and other exotic locations, all for science. Physicists weren't known for being swashbuckling types, but in 1932 at least, Arthur Compton came pretty close, although he would have strongly denied the label.

And in any case, he wouldn't be doing all the work and all the adventuring on his own. Beginning in 1931, he gathered an entire company of top-flight colleagues and laboratories to take part in what he called his "world survey of cosmic rays." He had little trouble in his recruiting efforts. "It was, of course, necessary in a program of this kind to get the extensive cooperation of a large number of physicists," he recalled. "Our invitations to join this work met with an enthusiastic response."[6]

Compton divided the globe into at first six, then later nine different geographical regions, selected to cover as broad a range of latitudes as possible. Compton would work in the Rocky Mountains, Canada, Hawaii, Switzerland, Australia, New Zealand, and Peru, accompanied on most of his travels by wife Betty and son Arthur. MIT colleague Ralph D. Bennett would go to Alaska, California, and Colorado; Compton's University of Chicago associate E. O. Wollan to Spitzbergen and Switzerland; and American Telephone & Telegraph research engineer Allen Carpe to Mount McKinley in Alaska.

Covering India, Singapore, Java, and Ceylon would be Compton's friend and colleague J. M. Benade of Punjab University, while Carnegie scientists under P. G. Ledig would work in Peru. South African measurements would be conducted under S. M. Naude at the University of Cape Town, and still more work by Dr. D. la Cour of the Danish Meteorological Survey in northern Greenland, near the Earth's magnetic pole. Compton even managed to enlist the aid of world-famous polar explorer Admiral Richard E. Byrd, convincing him to include one of Compton's fellow physicists with cosmic ray equipment on his next Antarctic expedition.

To ensure that all of the various readings across the planet would be reliable and consistent, each of the teams would use identical equipment, designed personally by Compton and accompanied by strict instructions.

The heart of the apparatus was an ionization chamber consisting of a 10 centimeter in diameter steel sphere triple-shielded by lead and bronze and pressurized with argon to 30 atmospheres. A radium source capsule provided a uniform calibration reference so that readings by different teams could be compared and any measurement errors corrected. Compton and his team nicknamed the device the "bomb" because of the striking similarity of the main piece, a round metal sphere, to a classic cartoon bomb, missing only the fuse sticking out of the top.[6]

"It was the scientific adventure of the age—and the largest group of scientific researchers that had ever been assembled on a common project," remarked Compton's younger son John in a 2006 reminiscence. "It was in the name of science, to be sure, but it was equally Indiana Jones—with all its risks."[7] Tragically, some of those risks would later prove fatal.

Pulling together and coordinating the efforts of so many people and resources around the world in a single dedicated scientific program in the early 1930s, long before the advent of communication and information technologies such as the internet, fiber-optic networks, and satellite links was a monumental accomplishment, but somehow Compton managed to make it all look relatively easy. A decade later, his experience in organizing and running his world cosmic ray survey would come in extremely handy when he became a major figure in the Manhattan Project.

For now, although Compton like almost any other competent physicist was quite aware of the potential of atomic energy for both good and ill, it was the power inherent in cosmic rays that was his chief concern. Or more precisely, he wanted to determine their true nature and, perhaps, their origins. "What is wanted is a survey of past studies, a careful verification of findings, a settling of one of the most contested and pressing problems in modern science," wrote *New York Times* science editor Waldemar Kaempffert.[8]

But as always, there was no way to write about cosmic rays without invoking the name of Robert Millikan. "It must be said for Millikan that every one of the questions that Compton will try to answer he has already answered himself," Kaempffert remarked. "He loses no opportunity to drive home his arguments in sledge-hammer fashion. He is the most diligent, the most forceful, perhaps the most thorough measurer of cosmic rays in the world today. The Compton experiments have been organized to settle the momentous issues that he has raised."[9]

So from the outset of his globe-girdling enterprise, before he or any of his colleagues had logged a single mile, Arthur Compton was already laboring under Millikan's long shadow, being cast willingly or not into the role of potential spoiler, competitor, rebel, a scientific David against Goliath. The stage was already being set for confrontation and controversy.

For Compton, it was not a matter of knocking anyone off any pedestal, just a dedicated scientific quest for the truth, wherever it might lead. Although he was reasonably certain that Millikan had some things wrong, Compton didn't yet have any clear ideas of what, how, or why; finding out was the entire point of the world survey.

Unfortunately, as Betty Compton later put it, "Mr. Millikan thought that that [cosmic rays] was his field, and he didn't feel that anyone else should undertake it."[10] And the Chief, sitting in Pasadena and taking in all the glowing press coverage of Compton's impending campaign, was already planning his counteroffensive.

"In February of 1932, Robert A. Millikan and Ira S. Bowen came into my laboratory to discuss the possibility of me joining them in a new series of measurements on cosmic rays," recalled H. Victor Neher, a Caltech physicist who had been busy studying electron scattering and developing a highly precise electrometer for the work. Neher was achieving some interesting results and was eager to continue, but Millikan thought otherwise. Never mind electron scattering now: Millikan wanted him to build "a self-recording ionization chamber and detector that was insensitive to vibration and tilt."[11]

Millikan knew that it was just the sort of thing that, appropriately modified, would be perfect for what he had in mind: a new series of cosmic ray measurements from airplanes, ships, and high-altitude balloons at various latitudes. If Arthur Compton was going to undertake a historic series of expeditions to probe the "secret of the cosmic ray," Robert Millikan was not about to sit still in his office in his "monastery of science," as he called Caltech's Norman Bridge Laboratory,[12] and let such an effort remain unchallenged.

Not that he would necessarily phrase it quite that way to a junior colleague such as Neher, who later wrote that "[Millikan] . . . went on to say that Arthur Compton had received a grant to make a world survey of cosmic-ray intensities at sea level as well as mountain altitudes. 'We welcome this kind of independent work,' he said. 'I'm sure we will all agree in time.'"[13]

A few days later, Neher presented his ideas for the instrument design to Millikan. The Chief approved immediately and proceeded to lay out a schedule almost as ambitious as Compton's, beginning with airplane flights from March Field in California that summer and proceeding to further flights from Washington State and northern Canada, then back to Seattle, from which further measurements would be taken by ship on the way back to Los Angeles. Then, while the elder Millikan remained in California, Neher, taking measurements all the while, would voyage by ship south to Peru for more airplane flights and mountain excursions, and eventually proceed by ship from Peru up to New York City to join Millikan at the annual AAAS conference at the end of 1932, where presumably he would join Millikan to present his results and final triumph over the doubters of birth cries and atom building.

Neher got quickly to work building his electroscope system with the help of several machinists. Millikan followed the work closely, never hesitating to offer suggestions in late-night consultations that would take place no matter what else Millikan had going on with his multitudinous scientific and administrative responsibilities.

"Appointments with him were often at night," Neher remembered. "He would say, 'I'll see you tonight in your room at 11 o'clock.' Sure enough, he would be there, and usually on time, and might stay until 2 o'clock in the morning. Sometimes he would come after a formal affair, still dressed in tuxedo."[14]

Always the hands-on experimentalist, Millikan had no hesitation in taking Neher's equipment out for a test drive—literally. To check parts of the apparatus for its sensitivity to vibration, the men loaded it up into the back seat of Millikan's 1928 Chevrolet for late-night spins around Pasadena. "He did the driving while I observed," wrote Neher, "changing the coupling between the vibrating car and the instrument by placing more or less human flesh in between the two."

To Neher, peering at quartz fibers through an eyepiece in a moving car, the system seemed to work just fine. But Millikan naturally had to see for himself. One night he was looking through the eyepiece as Neher took the wheel. Unfortunately, Neher wasn't used to driving Millikan's car. "I let out the clutch; it grabbed, and the car lurched forward. The eyepiece struck the bridge of [Millikan's] nose." Millikan came away from the experience with a minor scar on his nose for the rest of his life.[15]

Millikan's immediate reaction to the mishap is not recorded, but he apparently took it in stride. In any event, he wasn't about to fire Neher; he needed him too badly, not only to build his cosmic ray instruments but also to operate them as his globetrotting experimental proxy.

With the necessary equipment in hand, Millikan set about making all the logistical arrangements. Again, his scientific and academic prestige, reputation, and above all connections—especially his long-established relationships in the military hierarchy—would come in handy. It would probably have been quite difficult for almost any other scientist to get hold of military aircraft for something as relatively inconsequential as scientific research, but not for Robert Millikan.

As he had done some years earlier when he tapped his Signal Corps contacts for access to balloons for his Texas experiments, he now reached out to Lt. Colonel Henry "Hap" Arnold, commanding officer at Southern California's March Field, to request the use of high-altitude-capable bomber aircraft to carry his cosmic ray instruments aloft. Even without any prior acquaintance with Millikan, Arnold was already a soft touch for such a request, being an enthusiastic proponent and supporter of technical and scientific research efforts, particularly those involving aeronautics. While hunting down cosmic rays certainly had nothing to do with designing and building better and faster airplanes, Arnold was smart and intuitive enough to recognize the value of such work and was more than receptive to Millikan's entreaties.[16]

To cover all the bases and realizing that Arnold might have to get authorization from his Washington superiors to grant the request, Millikan also wrote to Major General B. D. Foulois, the Chief of the Air Corps at the War Department in Washington. In both letters, Millikan laid out the basic request: use of aircraft capable of reaching and maintaining altitudes of up to 25,000 for at least an hour "for the sake of finding the exact amount of the cosmic radiation at great altitudes, and its variation with altitude," using Caltech's "new and sensitive cosmic ray electroscopes." Millikan also explained that some preliminary calibration flights would be necessary in which the electroscope would need to be shielded with about 500 pounds of lead, which meant the aircraft chosen had to be capable of carrying such weight to the desired altitudes. But because the instruments could record their own readings, no observer or operator was necessary: "The only load that needs to be carried is the pilot and the recording outfit."[17]

Arnold responded promptly and positively. "We will be very glad to co-operate to the fullest possible extent," he wrote, noting that permission from Washington wasn't necessary and the only question would be reaching the required altitudes with the equipment weight. Arnold suggested the Curtiss Condor, a twin-engine biplane bomber. Unfortunately, Millikan would have to wait a month or so, since all of March Field's Condors were away at San Diego for bombing practice until August 1.[18] Foulois also gave his approval, forwarding Millikan's request to other involved personnel. Since Army Air Corps personnel weren't available at some of the other locations Millikan planned on using, Foulois suggested that Millikan reach out to National Guard authorities for those flights.

Throughout the rest of the summer, Millikan worked the mail and the phone to set things up. Aside from dealing with the military, he also arranged for sounding balloon flights with the U.S. Weather Bureau and began contacting Canadian airlines and military authorities to make arrangements for airplane flights for his northern expeditions.

Millikan plowed forward relentlessly through all the administrative red tape, bureaucratic delays and obstacles, and various and sundry equipment, financial, and miscellaneous annoyances that inevitably beset any such vast and complex project. Not long after the beginning of August, it seemed he was ready to finally get started chasing cosmic rays instead of generals and bureaucrats with the commencement of the March Field flights. That was until he was hit with, as he called it, "a streak of the worst luck that I have ever had in my life."[19]

By the end of July, Neher's new and improved electroscope system was finished and ready for final testing. It would be compared with and calibrated using Millikan's older instrument, the same one he had used for his 1925 mountain lake experiments. Because the precise characteristics and quirks of that instrument were so well known, it would provide a basis to gauge the performance of the new device. "By taking two values of ionization and comparing the two instruments, the constants of the new instrument can be determined in terms of the old, taken as a standard," Neher explained.[20]

After some initial tests in Pasadena, Neher and Millikan decided to try the new electroscope out in the field, back at Lake Arrowhead in California at the beginning of August. They set it up in a rowboat anchored in the middle of the lake and left it overnight. "When we looked out on the lake the next morning, there was no boat," recalled Neher. A squall had whipped through during the night and sunk it. Three days of dragging the lake produced a pile of old tires and other detritus, but no electroscope. Millikan was forced to ask Arnold for a delay in the flights for a few more weeks until a replacement instrument could be built. According to Neher, Arnold responded with good humor in a telegram: "Serves you right for playing with the Navy."[21]

Neher and his machinists managed to recreate the lost instrument in less than a month, and early in September it finally took to the skies, first from March Field and soon afterward from Cormorant Lake in Manitoba. Millikan and his crew soon discovered a somewhat annoying and unexpected byproduct of making cosmic ray measurements in aircraft. Along with the usual compensation that always had to be made for the unavoidable environmental background radiation, the experimenters had to adjust their calculations to allow for slight extraneous radiation from the radium paint in the cockpit instrument dials.

The measurements continued through the fall more or less as planned, with more airplane trips, sounding balloon flights, a trip to Pike's Peak, and the shipboard cruise from Seattle to Los Angeles. As the results and data so far collected were analyzed, Neher, for one, realized that the Chief was probably not going to be very happy.

"Even with some uncertainties about corrections for the airplane dials, it was quite obvious that there was a latitude effect of cosmic rays at airplane altitudes between southern California and Canada," he wrote.[22] Millikan, however, wasn't worried yet. There were still far more data to collect, including from Neher's South American odyssey, which would be the next and, Millikan trusted, the final nail in the coffin for both the latitude effect and any ideas about the "corpuscular" nature of cosmic rays.

Back in February, around the same time Victor Neher found himself being drafted into Robert Millikan's cosmic ray crusade, Arthur Compton was attempting a somewhat more subtle persuasion of his own.

"Mr. Gilbert of the Carnegie Institution writes me that you are undertaking a cosmic ray program for the coming year. . . . As you may know, the Carnegie Institution is also supporting some cosmic ray studies which I am undertaking. . . . It has occurred to me that in some of this work it would be valuable to cooperate in order that our results will be more closely compatible with each other," Compton wrote in a letter of February 14, 1932. It was certainly not a Valentine's Day message, but it was at least friendly, sent with good intentions. The recipient was Robert Millikan.

Compton had written the letter at least partly in response to a suggestion by Carnegie Institution officials, who were still concerned about duplication of effort between Compton and Millikan's separate programs. There was also an element of respect toward a senior colleague: "If there is any way in which in our travels to different countries we can aid your studies I hope you will let me know." Compton also mentioned the possibility of taking along a duplicate set of Millikan's instruments on one of his mountaintop jaunts, pointing out that "comparing our results on Mt. Evans with yours on Pike's Peak it appears that our measurements agree very closely with your own." And if Millikan was amenable to that idea, then perhaps he might be willing to take along some of Compton's equipment on an airplane flight near the north magnetic pole, to supplement Compton's own planned readings in the area.

Of course, Compton didn't write the letter merely out of political, or perhaps peacekeeping, motives. He was also concerned that he and Millikan risked wasting time inadvertently doing the same things twice, when instead they could work together to complement, rather than compete with, each other's work. "If any other type of cooperation suggests itself to you I should be glad if you would let me know."[23]

Compton's letter also mentioned a side issue that he wanted to bring to Millikan's attention. The maverick inventor Nikola Tesla has recently published a letter to the editor of the *New York Times* in response to the paper's recent exhaustive coverage of the cosmic ray subject. "Inasmuch as I discovered this wonderful phenomenon and investigated it long before others began their researches," Tesla wrote, "your readers may perhaps be interested in my own findings."

That launched a lengthy explanation in which Tesla claimed to have come up with and published on the concept of cosmic rays as far back as 1896. "But at that time scientific men were emphatically opposed to my theories. . . . It was only years later that the views I then propounded were gradually accepted." He described a series of experiments "facilitated through my invention of a novel form of vacuum tube." Interestingly, Tesla's ideas did involve the concept of "corpuscular" rather than photonic cosmic rays, specifically "inconceivably small particles animated by velocities vastly exceeding that of light," apparently emitted by "the sun and other heavenly bodies." Dismissive of probing the upper regions of the atmosphere with balloons and airplanes, Tesla believed "we will make much more rapid progress if those who are now taking interest in it will accept my theory and build further on this foundation, instead of embarking on useless errands in quest of mythical rays coming from nowhere."[24]

Tesla's notions were imaginative and fanciful as always. He failed to provide any concrete evidence for them in his letter, despite the fact that if true they would invalidate pretty much all of the foundations of physics for the last half century, including Maxwell's electromagnetic theory and Einstein's relativity. Compton had been sent a clipping of the letter by physicist Michael Pupin, who suggested that somebody, perhaps Compton, needed to set the record straight in the *New York Times*: "Tesla's letter cannot be ignored because, although he knows nothing of modern physics, there are many who think that he does, and it may do considerable harm if it is not pointed out that Tesla does not know what he is talking about."[25] Compton included a copy of the Tesla clipping and Pupin's note along with his letter to Millikan.

In passing along the matter to Millikan, Compton may have been showing deference to his more famous colleague or simply trying to pass the buck, but he was certainly more than respectful. "If an answer is to be made it would seem that you, rather than I should write it," Compton observed. "On the other hand my feeling is that there has already been too much newspaper publicity to cosmic rays and that this would merely be intensified by such correspondence." (Neither Compton nor Millikan would realize the ironic prescience of that statement until the end of the year.) "I should however be very glad to follow your lead in this matter."[26]

Millikan met Compton's offer of cooperation with studied indifference, not bothering to respond until almost a month later. He wrote that he would be glad to send along a Caltech instrument on one of Compton's polar flights, but because he and his team were mostly interested in spending the summer doing "upper air" measurements, it probably wouldn't be feasible to get a spare electroscope ready in time. He made no comment on including Compton's instruments on any of his own expeditions, or on the Tesla matter. Compton wrote back promptly, thanking him for the response and promising to keep Millikan informed of his plans in case things changed.

So ended a real opportunity for Millikan and Compton to work in concert rather than in conflict. Rather than two distinguished researchers pursuing two different but related avenues in the same subject, engaged at most in cordial, even friendly professional competition, they were now set on a course for an inevitable clash, one that because of their shared fame and public profile was going to be splattered across newspapers all over the world and especially in the United States.

There would be no more chances now to reach a *modus vivendi*, even if either man desired one. Millikan was now too busy working his vast network of connections, making travel plans, and building (or supervising the construction of) instruments. And Arthur Compton, just over a month after making his ill-fated pitch to Millikan, was getting ready to leave Chicago to begin his global odyssey.

8

GLOBETROTTING

On Friday, March 18, 1932, Arthur Compton, Betty Compton, and their thirteen-year-old son Arthur Alan stood on the steps of the Atchison, Topeka and Santa Fe Railroad train "Navajo," dressed in their finest traveling clothes, bidding farewell to friends on the platform at Chicago's Dearborn Station. They were headed for Los Angeles, where they would board the Los Angeles Steamship Company's flagship *City of Los Angeles* and sail for Honolulu. Compton's 20,000-mile cosmic ray quest had begun. "Cosmic Ray His Grail," proclaimed the *Philadelphia Inquirer* over Compton's photo.[1]

Arthur Alan was along to serve officially as a technical assistant, paid with funds from Compton's Carnegie grant. Betty would often do likewise, unofficially and unpaid, though she loved the work. For both, and also for Compton himself, it was a grand adventure, compared by son Arthur Alan to the explorations of Marco Polo, but now in the name of science.

A few days before boarding the train, Compton had given his equipment a final test run by taking it out for a spin in Washington Park, just west of the University of Chicago campus. All was well.[2]

Having already primed the public over the past months with coverage of Compton's plans, the press followed every move along the way. The reporting was not always completely objective. A number of newspapers,

8.1 Just before departing on his world travels in search of the cosmic ray, Arthur Compton tests out his equipment in Washington Park, near the University of Chicago campus. (Argonne National Laboratory, courtesy AIP Emilio Segrè Visual Archives)

particularly in their editorial commentary, were more than happy to stir the slowly gathering storm clouds of controversy.

Predictably, a telling example appeared in Millikan's hometown newspaper. In January, an editorial note in the *Los Angeles Times* about Compton's campaign was reasonably even-handed. "Although Dr. Millikan seems to have made a fairly exhaustive study of the cosmic rays, which he is credited with having discovered[3] . . . the Carnegie Institute deems the subject of such vast importance that it is assisting in financing an expedition to be headed by Prof. Arthur H. Compton of Chicago University."[4] The piece again briefly mentioned Millikan's pet "birth cries" and acknowledged the existence of alternative views, while noting, quite correctly, that "one skeptic says that 'about cosmic rays we have more opinions than facts.'" Still, the "general opinion of scientists" was that "radiation is the most important study in physics and that only by a more perfect knowledge of it can we solve the mystery of life and of the cosmos."[5]

Nothing too contentious there. But by the time Compton was actually getting started two months later, the paper's editorial take had become decidedly more overtly biased in favor of the hometown hero. "From Chicago comes the report that Dr. Arthur H. Compton . . . is about to start on his world tour of mountain peaks to 'trap' cosmic rays and find out where they originate," noted an editorial. The unnamed commentator sniffed that Compton's work "is merely an extension of the studies made of the rays by Dr. Robert A. Millikan, their discoverer." Again that erroneous credit, which was certainly common knowledge by now but remained irresistible to Millikan's legions of fans—especially in the Southern California press corps.

The editorial generously allowed that it was possible that Compton's efforts "may result in the accumulation of valuable data of a hitherto unknown character." However, "in the opinion of the Pasadena group of scientists it will be unnecessary to hunt for further proof of the fact that the rays . . . do not originate in our atmosphere. . . . Dr. Millikan . . . has brought what scores of scientists consider proof that the cosmic rays . . . must come from regions far more remote."

Having thus mischaracterized Compton's scientific objectives, the commentator concluded, "If these calculations are correct there would seem to be no need for Dr. Compton to travel his purposed 20,000 miles to the Antipodes or elsewhere either to prove or confute them. But there are a few physicists who doubt the theory of the spatial origin of the rays and Dr. Compton seems to be among them, else he would not spend six months in the most far-flung expedition in the history of physical science to conduct further observations."[6]

Compton's reaction to such commentaries, or whether in fact he had ever seen them, is unknown. But he would certainly have agreed that the work of his world survey would extend upon and complement the studies made by Millikan, which was one reason he had proposed collaboration with his more senior colleague. But he was likely far too busy to notice or care about press coverage. He had a lot of miles and a lot of work ahead of him.

Before sailing for Hawaii on March 26, Compton got word that two other parts of his program were getting under way as well: Dr. D. la Cour of the Danish Meteorological Survey had departed Denmark for Spitzbergen, and Allan Carpe was about to sail from Seattle for Alaska and Mount McKinley. So far, all was going smoothly according to plan.

"We stand at an epochal moment in physics," Compton told an interviewer in a lengthy *New York Times Magazine* profile that appeared the Sunday before he left Chicago. He was speaking not of his ambitious world survey beginning the end of that week, but of the accelerating flow of experimental and theoretical advances, including the recent discovery of the neutron by James Chadwick, that was increasingly revealing the details of atomic structure and now had physicists knocking on the door of determining the still-hidden nature of the atomic nucleus.

"It is my bet that the agency that is going to give the largest return on this problem is the cosmic ray," Compton opined. "Therefore I am centering my study on cosmic rays . . . by studying cosmic rays I expect to get a clue to the method of their generation, and therefore some knowledge of their generator, the nucleus."

As was *de rigueur* in nearly all science writing of the day, the reporter, George W. Gray, inevitably asked Compton about the prospects for atomic energy—and of course, God and the universe. "I am not speculating as to practical applications, though it should be obvious that the control of atomic energy would give man a power many times that of all the coal beds and waterfalls and all the tides and winds of the earth combined," Compton said. As to God, "the study of physics has changed my conception of the kind of god, but has strengthened my confidence in the reality of God . . . it seems to the nth degree improbable that such an intricate and interesting world could have ordered itself out of particles with a random character. I cannot conceive the thing happening wholly by accident."[7] Comforting Sunday supplement words for the public, to be sure, but also an honest statement of Compton's own beliefs.

Despite the impression that the average citizen might glean from the newspapers, however, there was far more going on in 1932 than the doings of Compton and Millikan. A new type of physics was being born, one that would become known as elementary particle physics and lead to profound revelations about the nature and structure of the cosmos. But Compton was right about several things: in vital ways, some direct and some indirect, the study of cosmic rays would provide the genetic material from which particle physics would be born. And for this and other reasons, 1932 would indeed become known as an epochal moment in physics, an *annus mirabilis*, a miracle year.

Back at the end of 1929, young physics Ph.D. candidate Carl D. Anderson met with his advisor, Robert Millikan. Anderson's thesis work involved using a cloud chamber to study the production of photoelectrons by x-rays, and he wanted Millikan's blessing to stay on at Caltech as a postdoctoral fellow to continue that work and also to learn more about quantum mechanics, a discipline for which Caltech had become a West Coast hotbed.

"I asked Millikan if this would be possible," Anderson recalled, "and he said, 'Well, it's a very poor idea. You've been here four years as an undergraduate and three years as a graduate student. You're getting very provincial.'"[8] Instead, Millikan recommended that Anderson continue his work at another institution under a National Research Council (NRC) fellowship.

The Chief had spoken, so Anderson duly applied for the NRC fellowship and approached Arthur Compton at the University of Chicago to inquire about going there. Compton readily agreed and Anderson began planning his move to Chicago as soon as he got his doctorate the following June, in anticipation of getting the fellowship.

He wouldn't make it there. "One day I received a call from Dr. Millikan asking me to see him in his office," said Anderson.[9] Millikan wanted him to stay on at Caltech for another year or so after all. The experiments Anderson had been conducting were just the thing that would come in handy for Millikan's cosmic ray program.

Historian Daniel J. Kevles explained, "Still sure that the rays consisted of photons, he also remained convinced that they were the birth cries of atoms forming in interstellar space. But Millikan had developed his atom-building hypothesis by inferring the energy of cosmic rays from the way they were absorbed in matter. To establish his hypothesis on a more solid experimental base, Millikan wanted to measure cosmic ray energies directly, and he thought the task might be accomplished with a cloud chamber set in a magnetic field."[10]

Anderson demurred as diplomatically as possible for a young Ph.D. just beginning his career. "I used all the arguments that he had previously made as to why I should not stay at Caltech. He replied that all these arguments were very valid and cogent, but that my chances of receiving an NRC Fellowship would be much greater after one more year at Caltech." Not incidentally, Anderson was also quite aware that Millikan was a prominent member of the NRC fellowship selection committee.

So, much as Victor Neher would later find himself similarly drafted into the Millikan cause, Anderson resigned himself. "Again, I seemed to have no choice in the matter, so without further ado, I began work on the design of the instrument [Millikan] had proposed for the cosmic ray studies."[11]

Anderson constructed an apparatus with an enormous electromagnet, strong enough to deflect high-energy cosmic rays. Because of the electrical power required, the gadget had to use the 600 kilowatt generator used by Caltech's Guggenheim Aeronautical Laboratory to power one of their wind tunnels. The heart of the instrument was a large cloud chamber, with a large square hole cut into one of the side magnets so that cloud tracks could be photographed. The whole apparatus was mounted on wheels so it could be readily pushed back and forth between the aeronautical lab for experimental runs and back home to the Bridge Laboratory for storage.

Anderson began work, spending hours and hours creating clouds and taking thousands of pictures of curving particle tracks. The results were startling and confusing, showing showers of both positive and negative particles but not all of them behaving as expected. By measuring the curvature of the particle tracks, their energy and charge could be determined, but some particles seemed to come from unexpected directions—from below, rather than from above.

Were the anomalous particles protons, electrons, or something else entirely? Anderson remembered that the possibility seemed to be "that these particles were either electrons (of negative charge) moving upward or some unknown lightweight particles of positive charge moving downward. In the spirit of scientific conservatism we tended at first toward the former interpretation, i.e. that these particles were upward-moving negative electrons."[12]

Millikan was skeptical. "This led to frequent, and at times somewhat heated, discussions between Professor Millikan and myself, in which he repeatedly pointed out that everyone knows that cosmic ray particles travel downward, and not upward, except in extremely rare instances, and that, therefore, these particles must be downward-moving protons," said Anderson.[13]

To settle the question, Anderson put a horizonal lead plate in the middle of the cloud chamber, which would cause a transiting particle to lose energy and curve in a distinct way that would prove once and for all whether it was negative or positive. On August 2, 1932, the proof arrived: a picture of a positive particle moving upward, but with the mass of an electron, not a proton. The twenty-six-year-old Anderson had discovered the positron, a brand-new and previously unknown particle, and the first example of antimatter, the existence of which had been predicted some years earlier by Paul Dirac.

Not everyone in the physics community was convinced at first, including such elder statesmen as Niels Bohr. It all seemed a little too unbelievable, a little too conveniently in line with Dirac's earlier theories. But further experiments by Anderson and his graduate student assistant Seth Neddermeyer, and separate observations by P. M. S. Blackett and G. P. S. Occhialini in Britain, who used the revolutionary idea of combining Bruno Rossi's coincidence circuit with a cloud chamber (Occhialini being a former disciple of Rossi), eventually confirmed the results. Four years later, Anderson would receive the Nobel Prize for discovering the positron, not incidentally providing yet another feather in the cap of Millikan's Caltech laboratory and his group of researchers.

As Arthur Compton had foreseen, the study of cosmic rays had led directly to a fundamental discovery. It would not be the last, as Anderson and Neddermeyer would later demonstrate. Unfortunately for Robert Millikan, it would also prove to be one of the first of many cracks beginning to appear in the edifice of his beloved ideas of atom building and birth cries.

"Although the 'atom-building' hypothesis did not appeal to Neddermeyer and me, it seemed to be very firmly fixed in Millikan's mind," Anderson observed.[14] It was going to take quite a lot yet to shake it loose—if that was even possible.

By April 2, 1932, Compton was in Hawaii on the island of Maui, heading up the side of the 9,300-foot Haleakala volcano on horseback, accompanied by his family and colleagues from the University of Hawaii. Loaded onto sturdy pack horses were about 400 pounds of equipment and supplies. The trip would take about four hours before they reached Olinda, the rest house at the summit, also known as the House of the Sun. It was a far cry from the Ryerson Laboratory in Chicago.

The excursion would not be without some problems. "[Compton] chose the horse he thought was the most stable one and put his precious battery box on the back of this horse, very carefully tying it on when the horse gave, as he said, the longest standing broad jump he had ever seen; and the box described a parabola and landed on a rock," Betty Compton later recalled. "Of course he went immediately to the rescue to see what had happened, and there were a couple of wires that got unsoldered." Always the resourceful experimentalist, Compton improvised a fix, using a candle flame to reattach the wires. "I can see him yet with a candle soldering those. And it worked."[15]

Compton and his crew spent all night atop the volcano crater rim, taking readings, occasionally taking time out to sleep and eat, and using candle lanterns both as sources of lighting and heating. After collecting about eighty separate measurements, they finally started back down the mountain the following afternoon. Three days later, they would depart for the next stop, Auckland, New Zealand.

Meanwhile, on April 16, Allen Carpe departed Seattle on the steamship *Alaska* en route north to Mount McKinley. Aside from being an engineer with AT&T, Carpe was known as an expert mountaineer with a great deal of experience not only in Alaska but also elsewhere in the Canadian Rockies—a man with both the practical skills and the technical acumen to lead the expedition and make the necessary measurements using the Compton electroscope.

He would never return home. Carpe and his colleague Theodore Koven would be lost on Muldrow Glacier, apparently both falling into a crevasse and dying of exposure. Carpe's data notebook, though not his body, was eventually found; Koven's remains were finally recovered and returned home for proper burial in August. On June 2, Compton's MIT colleague Ralph D. Bennett was tapped to take over and complete the Alaskan part of the cosmic ray survey, with a marked change in strategy: "Being scientists first and mountain climbers secondarily, the Carnegie Institute expedition will use the ore buckets and cables of mining companies to scale peaks in Alaska this summer."[16]

A guide on Dr. Naude's South African expedition would also die during the year, though under less exotic circumstances. The Carnegie Institution may have been providing the funds for all the exotic travels, equipment, and various other expenses of the Compton campaign, but tragically, these three men paid a far heavier and unexpected price for the sake of scientific discovery.

On the other side of the planet, Arthur Compton was likely not yet aware of the tragedy that had just occurred in Alaska, busily at work in Australia and New Zealand. In Pasadena, Millikan was making his own preparations for his own slightly more modest crusade as Neher continued work on the new electroscope. He also found himself receiving a bit of scientific affirmation from his colleague Fritz Zwicky that, as *Los Angeles Times* reporter Ransome Sutton put it, "the universe is not blowing up."[17]

Before leaving for a honeymoon trip to Switzerland, Zwicky, "who ranks with Einstein, De Sitter, Tolman and Lemaitre in cosmological investigations," lectured about cosmic rays at Caltech. He maintained that among other things, the measured energies of cosmic rays seemed to indicate an intergalactic origin, providing another possible explanation for the red shift observations pointing toward an expanding universe that would eventually come to a dark and dismal end. "Everybody heaved a sigh of relief: the universe may not be exploding after all. It may even be static, in which case Einstein will be sorry he changed his mind," Sutton wrote.[18]

The reporter also noted, "One of the intensely interested listeners was Dr. Millikan. Hearing his theory of the origin of the cosmic rays thus ably maintained, he seemed to be saying: 'I told you so.'"[19]

The *Los Angeles Times* never missed a chance to boost the name and visibility of Southern California's most eminent resident scientist, no matter how trivial the matter. A few days after the Zwicky report, which at least was a legitimate news story, the paper felt obliged to recount "a humorous poem on cosmic rays" published in the Caltech student newspaper: "What guides the planets through the skies? Why does the universe vary in size? What great, deep-lying emotion causes astronomers with great devotion, straining their eyesight in vain hopes of seeing something through their telescopes, night long sit fast and idly gaze? No riddle at all, just cosmic rays." Along with the now-obligatory claim that "Dr. Millikan's research brought the mysterious rays to the attention of the world," the item noted that although Millikan's youngest son was one of the student paper's editors, "it is asserted" that he didn't actually pen the poem.[20]

Millikan may have been feeling confident and assured (if not necessarily amused by his son's ostensible literary talents), but he would soon have cause to feel otherwise. Compton's work was already beginning to pay dividends, and they were not the sort of returns that supported the cherished philosophies of Robert Millikan.

Barely a month into the world survey, and with many more measurements from many more geographical areas yet to come, Compton was already feeling confident enough to present his first preliminary results in the scientific literature. In a brief letter to the editor of *Physical Review*, a well-established venue for quickly announcing important new findings to the scientific community, Compton described "definite differences in the intensity of the cosmic rays at different latitudes," with a "uniform variation with latitude, showing a minimum at or near the equator, and increasing intensity toward the north and south poles."[21] He briefly described the design of his instrument and provided a short table of readings, including his 1931 measurements from Colorado and the Alps up to his most recent data from Australia and New Zealand.

He also included a graph comparing his readings with those of Millikan and his collaborators, noting that while both sets of measurements at similar latitudes were largely in agreement, "the difference with latitude is greater for the less penetrating [i.e., less energetic] rays."[22]

Compton was not completely sure just yet what it all meant, but so far at least, "the variation of the cosmic rays with latitude which these measurements show is of just the kind to be expected if the rays consist of electrically charged particles which are deflected by the earth's magnetic field, the less penetrating rays being the more strongly affected."[23]

Compton's letter to *Physical Review* was dated May 7, 1932, and would not actually appear in the journal until its July 1 issue. Before that happened, however, Compton, along with Betty, Arthur Alan, and the rest of his party would complete further measurements in Panama and Peru. His

8.2 Ralph D. Bennett, one of Compton's colleagues on the world cosmic ray survey, takes readings in the mountains of California. (AIP Emilio Segrè Visual Archives)

data would continue to demonstrate a latitude effect as well as an increasing intensity in cosmic rays with higher altitudes.

While Compton cruised the seas and scaled mountains with his trusty electroscope, the press reports kept coming, some more sensationalist than others. An Associated Press report from Panama City observed that Compton "said today recent investigations by his cosmic ray expedition showed a ray which penetrates 18 feet of lead. It is definitely affected by the earth's magnetic field, he said, and added this indicated it was not light, but a stream of electrical particles."[24] Added to this a couple of days later came the editorial comment: "A physicist at the University of Chicago is said to have perfected a cosmic ray which penetrates eighteen feet of solid lead. Well, well, well. It may be possible that the light may get through some of our Washington statesmen's heads after all. There is hope!"[25]

When Compton's *Physical Review* report was published, however, the journalistic coverage took on a decidedly more serious tone. "New Data

Challenge Cosmic Ray Theory," proclaimed the *New York Times*. "Compton Reports Findings That Undermine Millikan Hypothesis Creation Is Still Going On." The latitude effect shown in Compton's data removed "a basic prop" of Millikan's theory "that the process of creation was still going on." Instead, said the *New York Times*, Compton's findings "would eliminate, for the present, all traces of experimental evidence for the hypothesis of continuing creation."[26]

Not that either Compton or Millikan were planning it that way, but the lines of battle were being clearly laid out, not merely in the scientific press but also in the public media. The day after the *New York Times* report, Milllikan's impending campaign of airplane and sounding balloon flights was announced.

Ever the canny operator, Millikan was playing it close to the vest with regard to the press, telling the *Los Angeles Times*: "Until we obtain results I do not care to discuss the projected research in the upper air in detail, but we hope the experiments will be more significant than any conducted on the hemisphere." Such as, one might assume, those of Compton.

As usual, after noting that Compton "has obtained results which are in conflict with research data of the local scientist [Millikan]," the article from the *Los Angeles Times* couldn't resist observing that nothing less than the fate of the universe was apparently at stake. If Millikan's theory was correct, "The faith of those who believe the universe constantly is being re-created and will never die will be strengthened." On the other hand, if cosmic rays proved to be not "birth cries" but "death wails," then "Sir James Jeans and other jovial scientific pessimists will find the greatest stumbling block to their theory that creation is doomed for a remote but certain incineration, removed."[27] For the time being at least, Millikan seemed content to let the *Los Angeles Times* serve as his unofficial cheerleader and stalwart defender.

Further dispatches later in July from Compton, now deep into his two-week Peruvian odyssey, seemed to provide more support to the jovial scientific pessimists. Readings from sea level all the way to the 19,000-foot summit of El Misti near Arequipa continued to confirm the developing picture of a magnetic latitude effect.

"Both sets of measurements agreed, showing increasingly rapid increase in intensity with the altitude," Compton reported in a press statement.

"This is contrary to some of the balloon measurements made eight years ago by Prof. Robert Millikan, but confirms the results obtained by Prof. Auguste Piccard on his recent balloon flight into the stratosphere."

Although his survey was far from completed and he had not yet officially published any results aside from the brief *Physical Review* letter, Compton couldn't resist adding that his measurements "support the theory that cosmic rays are electrical in character. Prof. Millikan's balloon experiments have been considered to give strong evidence favoring the alternative view that cosmic rays are not electrical, but similar to light."

The difference in the way these remarks were presented respectively in Millikan and Compton's hometown newspapers is intriguing. The report by C. N. Griffis of the *Chicago Tribune* bore the headline "Cosmic Rays Get Stronger at High Levels: Compton." The *Los Angeles Times* version published on the same day, nearly verbatim and marked "Exclusive" without a byline, is titled "Millikan's Ray Theory Rapped." Also interesting is the fact that while the *New York Times* likewise paraphrased the *Chicago Tribune* report, they credited the Chicago paper, unlike the *Los Angeles Times*. Millikan's journalistic champions were on the job.[28]

If anyone had any doubts on that score, there was reporter Bailey Millard's column in the *Los Angeles Times* several days later, titled "Are Cosmic Rays Merely Electrons?" Again making sure to point out the cost of the Compton effort, Millard contended that most of what Compton had so far announced had either already been predicted previously by Millikan or was insignificant. The idea of "electrical particles" being affected by the earth's magnetic poles "as streaming electrons may be diverted to one side by laboratory magnets" had already been suggested by other physicists and was inconsistent with Millikan's views. Of Compton's ideas, however, "what this theory lacks is proof." Even if the *New York Times* thought otherwise, Dr. Millikan was not about to abandon his theory. "It is going to be rather difficult to prove that cosmic rays are not electrons, but I will back the Pasadena savant to prove it if he sets out to do so."[29]

Millikan continued to hold himself above it all, playing his accustomed role as eminent savant. "Mankind, through the facilities thus far provided by science, now faces a century in which it knows, if it will act rationally, instead of irrationally, that all its physical wants can be more than met," he intoned in an address to 300 delegates to the International

Recreation Congress at the Biltmore Hotel in Los Angeles.[30] It was yet another of the endless round of formal events in the social and professional whirlwind that filled the evenings of a Nobel Prize-winning scientific authority.

By August 9, Compton arrived at Vera Cruz, Mexico, on the steamer *Orizaba* for yet another leg of his journey, a ten-day Mexican mountain survey, before moving northward to Canada. As previously noted, by this time, Millikan and Neher were testing out and losing their new electroscope in the depths of Lake Arrowhead. Neher remained behind while Millikan returned to Pasadena in search of a diver to recover the instruments.

On the other hemisphere of the planet, other adventurous researchers were preparing to venture forth into the far reaches of the sky, not to break records, but for science.

9

SKYSCRAPING

Balloons were very much on the minds of many people in the summer of 1932. Arthur Compton may have been spending his time riding horses up mountainsides, but Millikan was busily querying balloon manufacturers and setting up sounding balloon flights across North America under the auspices of the U.S. Weather Bureau. And in Europe, two other scientists were launching balloons in the name of science, one remaining safely on terra firma, the other riding along.

Professor Erich Regener had been among the early European researchers who began studying cosmic rays in the 1920s, inspired both by Hess's work and later by Millikan and Cameron's experiments. He began with a series of underwater experiments in 1928, sinking an ingeniously designed ionization chamber almost 800 feet into the depths of Lake Constance. Spurred by Auguste Piccard's first stratospheric flight in 1931, Regener began concentrating his attentions skyward to working with balloons. "[Piccard's] flight stimulated of course Regener to enter the competition between manned and pilot balloons and to speed up the efforts of getting ahead with his instrument," recalled his colleague Georg Pfotzer.[1]

After a series of unsuccessful attempts, Regener finally succeeded on August 12, 1932, in launching two six-and-a-half-foot diameter balloons attached to a small gondola packed with photographic plates and recording devices from the courtyard of the Stuttgart Institute of Technology.

Small, light, and impeccably designed, Regener's craft easily beat Piccard's manned altitude flight record, rising to an altitude of about seventeen miles.

Like its most vocal and visible member Robert Millikan, Regener had long been a firm adherent of the cosmic-rays-as-photons camp, but after safely recovering his instruments and examining the data they had collected, he found himself beginning to change his mind. As Compton's mountaintop results were showing, Regener also discovered that the cosmic ray flux began to diminish above a certain altitude, apparently around 39,000 feet—more evidence for a particle, not photonic, nature.[2]

Meanwhile, Compton settled into his Mexican work atop the Nevada de Toluca volcano at 15,000 feet. This time, at least, he didn't have to deal with pack horses and mountain trails, since a paved highway conveniently led up to the summit. He even enjoyed the use of a field house with beds and a telegraph line to the outside world, courtesy of the Mexican state governor.[3]

Far from Mexico but considerably closer to Stuttgart, Auguste Piccard and his assistant Max Cosyns soared aloft toward the stratosphere on August 18 in a gondola redesigned with lessons learned from the previous year's flight. This time things went far more smoothly, and Piccard and Cosyns achieved a final altitude of just over ten and a half miles—a new manned flight record. Calling the flight "a profitable trip," Piccard immediately began laying plans for yet another excursion, this time from Hudson Bay, Canada. "Our purpose will be to complete the study of cosmic rays at a point where the lines of magnetic force penetrate the stratosphere," he said. Millikan and Compton, of course, had much the same thing in mind, though in their case the measurements wouldn't involve either of them shivering in the swaying gondola of a balloon ten miles above the ground.

The summer of 1932 had become perhaps the busiest and certainly most far-flung period of cosmic ray research activity since Victor Hess's initial discovery twenty years earlier. What had begun as a somewhat tentative scientific enterprise, almost an afterthought or backwater of physics in the shadow of everything else that was going on with the revolutions of relativity and quantum mechanics, was now a mature and respectable discipline of recognized importance and a significance that transcended physics to involve astronomy, chemistry, cosmology, religion, and popular culture.

Anyone who was weary of hearing about cosmic rays was out of luck. Even with the Great Depression, an election, and political turmoil in Europe as distractions, cosmic rays were not about to disappear from the American zeitgeist anytime soon. In fact, they were just getting started. Between Piccard, Compton, Millikan, and Regener, remarked *TIME* magazine, "the quest for the cosmic ray was a four-ring circus."[4]

As Piccard was recovering from his latest aeronautical triumph and Compton was wrapping up work in Mexico, Millikan was taking his own new instrument out on the road and up to the skies. Under the command of Lieutenant Charles Howard of the Eleventh Bombing Squadron, the Condors of the Army Air Corps spent several days flying Millikan electroscopes to graduated altitude steps for forty-five minutes each, beginning at 10,000 feet, moving to 15,000, 19,000, and then in thousand-foot increments until reaching a maximum height of about 23,000 feet. The flights began each day at 6:00 a.m. and concluded about twelve hours later, just before darkness fell. Millikan, Neher, and Bowen took turns, one of them flying along while another stayed on the ground taking readings from another electroscope for comparison.

The press followed closely, of course, making much of the fact that Millikan's ingenious new electroscope was "ten times as sensitive as any that he hitherto has known" and that he and his two assistants, Neher and Ira S. Bowen, had "the air forces of two nations at their command." In case anyone had missed the point from all the previous coverage of Compton's roamings and Millikan's theories, and "despite Dr. Millikan's refusal to discuss the question," it was a foregone conclusion that the primary objective of the enterprise was to "settle the friendly differences" between Millikan and Compton.

After the March Field flights wrapped up, Millikan and company set out from Pasadena on their own international adventure on September 1, though one rather less ambitious and much shorter than the journey Compton was then completing. Covering his departure from the Santa Fe railroad station, reporters couldn't resist asking Millikan whether he had any thoughts about Compton's latest reported results.

But Millikan continued to play it coy. "That will come up later," he told the press, reportedly with a twinkle in his eye. Said the papers, "If the Pasadena scientist fears his theories are in danger, he gave no sign of it as he greeted friends who bade him 'bon voyage.'"

Perhaps the twinkle in Millikan's eye came from the knowledge that, at least according to the *Los Angeles Times*, "The consensus [sic] in local scientific circles is that Dr. Compton, while one of the greatest physicists, is a newcomer in cosmic ray investigation, and thus more likely to stumble than is Dr. Millikan."[5]

Whoever might stumble first, it was clear that more was at stake than the answer to an arcane scientific issue. Nothing less than immortality and endless creation versus chaos and utter oblivion were riding on the outcome. While Millikan was setting off for Winnipeg and points north, Compton was already even farther north, doing his best not to stumble, and collecting still more data that threatened to stifle the twinkle in Millikan's eye.

Somewhere inside the Arctic Circle about 350 miles from the Earth's north magnetic pole, Arthur Compton sent a wire back home to Chicago, around the same time that Millikan was leaving California. In an exotic place, he had been observing an exotic phenomenon.

The main question with regard to cosmic rays, along with their ultimate origins, was whether they were waves (i.e., photons), or particles (i.e., corpuscular), or composed of some variety of subatomic particles, with electrons being considered the chief candidate in mid-1932. The existence or nonexistence of a latitude effect would be the decisive factor in answering that question. But there were other side issues related to the big questions, such as the role of the sun in cosmic rays. Did some rays come from the sun? One way to find out was to determine what happened when the sun was absent from the sky, such as at night or during a solar eclipse.

On Wednesday, August 31, 1932, as Millikan's Condors were probing the Southern California skies, Compton was 100 miles north of the Arctic Circle, working his electroscope during a solar eclipse that reached 95 percent totality. His readings showed no effect at all from the eclipse, but a definite magnetic effect caused by the proximity of the planetary magnetic pole. This, noted a University of Chicago announcement, was confirmation "that cosmic rays are electrons, rather than extremely short-length magnetic vibrations."[6]

While Millikan was flying in California and Compton was freezing in the Arctic, P. M. S. Blackett and G. P. S. Occhialini were announcing the first pictures from their cloud chamber/coincidence circuit cosmic-ray detector, work that would confirm Carl Anderson's recent discovery of the positron. Added to the news in the *Los Angeles Times* was the aside that similar pictures of cosmic ray tracks "have been photographed by Dr. Robert A. Millikan of California Institute of Technology. He believes the rays are photons, or 'bullets of light,' while other scientists insist they must be electrons." The piece also included the familiar references to Millikan's "immortal universe" theory and James Jeans's opposing "dying universe" ideas.[7]

A few days later on September 6, Millikan, Neher, and Bowen arrived at The Pas in Manitoba for their own high-altitude Arctic explorations. "The importance of this investigation is more momentous and of greater import than the Piccard excursion into the stratosphere," Millikan told reporters. "It will undoubtedly mark a new phase in the study of the cosmic ray." Apparently, Compton's excursions were too inconsequential to comment upon directly, although he did note that the current work "is concerned with the effect of the magnetic pole on the cosmic ray . . . from our tests here we will find out if latitude has any effect on the rays."[8]

Millikan and his crew immediately set forth to Cormorant Lake fifty miles north. The next day they would head skyward for a series of stratospheric readings to compare with those done a week earlier at March Field. Royal Canadian Air Force pilot R. A. Gordon and John Fortey would fly the planes while Millikan and company took the measurements.

There would be a little excitement with the second flight, on which Millikan stayed behind. Gordon was late getting back to base. As he later explained, he and Fortey had to make a forced landing on a lake about 300 miles north and then became temporarily stuck on a sandbar, but after some quick repairs managed to get airborne and safely home again, not incidentally breaking the then-standing Canadian flight altitude record of 22,000 feet. Millikan was reported as "jubilant" at their return, not only because the pilots were safe, but no doubt also because he hadn't lost another valuable electroscope in another lake, not to mention the collected data.[9]

Meanwhile, as Millikan's student Carl Anderson was preparing to publicly announce the existence of the positron in *Science* on September 8, one of Millikan's cosmological debate partners, Sir Arthur Eddington, was speaking about the expanding universe to a congress of the International Astronomical Union at the Massachusetts Institute of Technology. With a crowd of more than two thousand people spread over two halls, it was at the time the largest audience that had ever attended a scientific lecture in the Boston area, with hundreds turned away for lack of room. The distinguished British scientist could certainly draw a crowd.

Introduced by Karl T. Compton, MIT president and Arthur Compton's older brother, Eddington compared the expanding universe to an inflating balloon, explaining that as dots on the balloon's surface would move away from each other as the balloon inflated, so did astronomers observe the distant galaxies receding from us and each other with the expansion of the universe. Just as theory predicted, "The remote spiral nebulae appear to be running away from us and the velocity of recession increases in proportion to the distance," as Edwin Hubble's work had demonstrated.[10]

Of course, anyone speaking about universal theories involving the birth and death of all creation couldn't avoid dealing with the looming figure of Robert Millikan. "Sir Arthur, in discussing the journey of light in a spherical universe, raised the question concerning the correctness of Dr. Robert A. Millikan's theories on the sources of cosmic radiation," observed the Associated Press. Eddington thought that rather than being continuously generated as Millikan's ideas would have it, most of the cosmic rays observed from Earth were actually far older, possibly dating as far back as the beginning of the universe's expansion itself. Nature, in fact, might be "playing a joke" on scientists who held that cosmic rays were of comparatively recent origin. As the *Boston Globe* put it, "Sir Arthur believes that Prof. Robert A. Millikan of California has much to learn yet about cosmic rays and where these rays come from."[11]

Farther north, Millikan was continuing his efforts to learn more about cosmic rays. He had wrapped up his business with the Royal Canadian Air Force, having obtained what he called "very fine records," and was about to stop off in Winnipeg for the weekend. He and Bowen would then continue south to North Dakota for some sounding balloon work while Neher took an instrument to Pike's Peak.[12] After meeting in Colorado, the team would proceed to Spokane in Washington State for more high-altitude measurements from army bomber aircraft while Bowen went down to Texas to make some further sounding balloon readings. Finally, Millikan and Neher would head back home from Seattle on the sea cruise to Los Angeles.

Before leaving Canada, however, Millikan took the time to deliver a lecture *cum* sermon to an audience *cum* congregation at Winnipeg's Westminster United Church. "The possessor of a virile Christian faith, Dr. Millikan has taken a leading place among the scientists of the continent in bringing about better relations between the forces of science and religion," noted the *New York Times*. "He is a member of the Congregational Church at his home in Pasadena, Cal. The church here was jammed with eager listeners." As he had done many times before both in writing and in speeches, he declared that there was no essential conflict between science and religion, and indeed many of the younger generation of physicists were also religious. According to Millikan, "Some of the most progressive and penetrating of modern thought has been done by theological groups."[13]

However secure Millikan may have been in his own personal spiritual certainties, he would soon suffer more earthly disturbances. The balloon work in North Dakota went badly; the first launch attempt on September 14 was thwarted by high winds that sent one balloon up without its instruments and ripped two other balloons to shreds. Two days later, another balloon flight developed a leak and came down early without reaching altitude, though the instruments were recovered. At least one other flight was made in a hybrid configuration, launching a balloon and electroscope from an airplane at about 21,000 feet. Uncertain of how much data he had managed to gather, and with some instruments lost and yet to be returned by whoever happened to find them on the Dakota plains, Millikan packed up and moved on to join Neher in Colorado.[14]

On September 29, sailing on the Admiral Line's *Emma Alexander*, a weary Millikan and Neher made it back to Los Angeles. Bowen was wrapping up balloon flights in Texas. Now would begin the laborious process of collating and analyzing all the data from all the locations. Prodded by reporters, Millikan maintained a scientific nonchalance. It would take time before anything definite could be said about results or the actual success of the campaign, he said. Nor would he say anything specific about the balloon troubles in North Dakota or Piccard's lofty adventures in Europe, except that he was confident that whatever data Piccard had managed to collect would be transcended by those captured by his own highly sensitive electroscopes.[15]

As to the doings of Arthur Compton, Millikan had even less to offer, apparently having decided to hold his cards close to his chest until there was more news on that particular front. That, however, would change very soon and very quickly. Compton was returning from the far north, and he had quite a lot to talk about, most of which directly challenged and contradicted Pasadena's "relentless hunter of the cosmic ray."

10

THROWING THE GAUNTLET

Robert Millikan persisted in telling the press that the results of his recent two months of cosmic ray observations across North America would not be known for months. Perhaps he was simply being a prudent, careful scientist who makes no definitive assertions until sure of his ground and his data. Or perhaps he was beginning to feel somewhat unsure of his ground under the unrelenting and unprecedented assaults he had been undergoing from all sides in recent months, among them from Jeans, Eddington, Rossi, and now most publicly, Arthur Compton, none of whom seemed at all interested in deferring in any way to the legendary Millikan.

Fortified with a new confidence about his work from his epic journeys across the planet, Compton was already out there swinging, even before he made it all the way back home to Chicago. And the press was delighted, especially when he made statements such as, "My work on the Arctic barren lands shows the cosmic ray is an electron, not a wave, as Dr. Millikan believes." No ambiguity there: that was about as direct a challenge as one scientist could present to another.

Compton began going public with his findings as soon as he could, beginning with his stop at The Pas in Manitoba on his way home, the same place Millikan had passed through just about a week earlier. Compton was full of juicy remarks. "The eclipse had no effect upon the cosmic ray," he proclaimed. But the proximity to the magnetic pole definitely did. "Obviously if the north magnetic pole has any effect on the rays, they must be electrical in nature instead of a wave, as Dr. Millikan contends."

And the latitude effect shown by the polar vs. equatorial measurements was also quite real. "The difference shown by my experiments will be a severe blow to Dr. Millikan," Compton said, adding that "it is probable" that Millikan's recent high-altitude flights in California and Canada "will confirm my work." Just in case anyone still wasn't getting the point, a subheadline of the *New York Times* report spelled it out explicitly, and in capital letters: "HOLDS MILLIKAN IS WRONG." The *Los Angeles Times* made it even more personal in their headline: "MILLIKAN'S RAY THEORY DENIED."[1]

That was only the beginning. Compton continued to make further statements at various stops on his way home, coinciding with the publication of another short letter in *Physical Review* about the particularly powerful cosmic rays detected at high altitudes on South American mountaintops. The *Los Angeles Times* headline was particularly lurid if startlingly inaccurate: "Terrific Atomic Blasts Discovered on Mountain." Again, Compton emphasized that the work provided more evidence that the rays were electrons rather than photons.[2]

As if to stoke the rising controversy still further, the *New York Times* also published a summary authored by Compton himself in which he again emphasized that his worldwide experiments, along with the recent findings of Piccard's flight and Regener's sounding balloon work, proved that the latitude effect was real, that the rays were electrons, and that there was a continuous rise in intensity with increasing altitude. He made it clear where he stood: "If the cosmic rays are streams of electrons instead of waves, it means that theories such as those of Professor Millikan, who has assumed that they are waves emitted on the formation of atomic nuclei, or Professor Jeans, who supposes they are waves produced by the destruction of atomic nuclei, will need to be radically changed."[3] Millikan and Jeans were apparently both wrong.

Compton, however, was merely firing some opening shots across Millikan's bow. In his press statements, he made clear that his final results would only be presented in their entirety at the American Association for the Advancement of Science meeting at the end of the year, in Atlantic City. Millikan continued to remain aloof and silent back in Pasadena, perhaps with the idea—or hope—of allowing Compton to hoist himself on his own petard. His old friend and colleague Frank Jewett, president of Bell Laboratories, thought that was a good idea. "I have been mildly interested and somewhat amused at the cosmic ray discussion in the public press this summer (more properly I should say in the cosmic ray attack since apparently you have not participated to any great extent)," he wrote to Millikan in late September. "Whatever the merits of Compton's position, I have a feeling that he has not done himself any good by rushing into print in the daily press. I should think the wiser thing to do would be to present his data, theories, and conclusions formed, through the Physical Society."[4] Millikan undoubtedly agreed.

Though not yet anywhere near a full boil, matters were definitely simmering in the press. The idea of two Nobel Prize-winning scientists at odds over the nature and fate of the universe was simply too appealing a prospect to ignore, and too ripe for all sorts of commentary. The *New York Times* provided two prime examples on its editorial page the day after publishing Compton's short article. A piece titled "Cosmic Ray Romancing" compared the various efforts and theories of Millikan, Compton, Hess, Regener, and Jeans and simply threw up its hands, implying that in the end it was all a matter of personalities and opinions.

"The physicist is still the medicine-man from whom he is descended," said the unnamed author, comparing electroscopes and ionization chambers and other instruments to "wands and totem poles" and equations to "incantations that make us believe we know more than we actually do." In the end, supposedly, no one really knew what they were talking about. "That we are actually dealing with something like wish-fulfillments in the cosmic rays is evidenced by the results obtained," the article continued. "Here is Millikan convincing himself that the cosmic rays prove that the universe is self-perpetuating. And Compton, adopting precisely the same methods, reaches the conclusion that the rays are only electrons swerving to the Poles because the earth is a great spinning magnet. . . . There is no positive answer. We simply try to reconcile what the instruments indicate with our hopes and beliefs and imagine we understand the cosmos."[5]

A companion editorial complained that since scientists usually seemed to insist on having things both ways, "It is a relief to read that Professor Compton is back from studying cosmic rays in the Arctic region with the definite report that Professor Millikan is wrong. . . . It is a relief to find that when two men in the high realms of science hold opposite views one of them is right and the other is wrong. Hitherto the public has had to get used to the idea that when two great physicists differ radically about something in the universe the answer is that both men are right."

The piece noted recent ideas and theories of wave-particle duality, a contracting vs. expanding universe, and the like, complaining that scientists could always manage to come up with a mathematical formula to support any idea. "Obviously it is a delightful world in which you can have the coffee simultaneously hot and iced out of the same cup," which made the Millikan-Compton differences refreshing, as "a relief to find black as the opposite of white and right as the counterpart of wrong."[6] The journal *Science* also republished both editorials without comment in its "Quotations" column, apparently as an example to its scientific readership of the sort of outrageous things being bandied about in the popular press.[7]

Meanwhile, the *Los Angeles Times* editors engaged in more Millikan boosterism, extolling Anderson's recent announcement of the positron or "positive electron." Perhaps it hadn't been Millikan himself who had made the discovery, but Anderson was certainly an associate of the Chief, and this latest announcement was yet another triumph. "Wonders have been done with the atom by the Millikan school of physicists—wonders that have made the whole scientific world sit up and take notice," gushed the editors. "And still another chapter will be added to the story of these activities when Dr. Millikan returns from Canada and examines the electroscopes he has sent up to a height of 22,000 feet to solve the riddle of what effect, if any, earth's magnetic force has upon the cosmic rays."[8] Apparently the authors had not been reading the recent issues of the *New York Times*. In another tidbit a few days later, however, they did make note of the recent bomber flights made "to help Dr. Millikan catch the frisky cosmic ray."[9]

Frisky cosmic rays indeed seemed to be everywhere in the fall of 1932. Even the sports pages were not free of their presence, at least in the *Los Angeles Times*, where Caltech football coach Fox Stanton, in a column giving his particular recipe for football success, observed, "Everybody and everything alive is hunting for something. . . . Even at this moment one party of physicists are hunting for a cosmic ray which, when it is safely ensconced in a test tube, they will transport with much ado to a laboratory, where it will be determined whether a cosmic ray is something to be turned into dollars or to be swallowed alive. . . . What has this to do with football? . . . Not a thing in the world!"[10] At least Stanton didn't mention Millikan by name or say anything about universal birth cries.

The Millikan school of physicists may have been laying low for the moment, but they were far from idle. Preparations were being made for Victor Neher's journey southward, where presumably he would gather new data that might refute the Compton measurements. The four weeks or so after the return of Millikan and his crew to California "was spent in making calibrations and making a new quartz system" for the electroscope Neher would be taking along. "It seemed foolish to leave with just one instrument, with no preparation for failure, Neher remembered."[11]

Time was a major factor. No matter what, Millikan wanted the Peruvian expedition completed and Neher and the data returned in time for the AAAS meeting in December. The pressing schedule led to some inevitable cutting of corners in the preparation of the instruments. The shortcuts seemed minor at the time, but they would eventually have major and embarrassing consequences for Millikan.

Arthur Compton finally made it back home to Chicago at the beginning of October with "a trunkful of new data on the cosmic rays," as a University of Chicago press release termed it.[12] The main conclusions from his recent peregrinations were again emphasized: the latitude effect, the continuous increased intensity with altitude.

Compton would now spend some time unpacking and catching up on work, while analyzing his new cosmic ray data and waiting for results to come in from his other still-ongoing expeditions in Spitzbergen, Norway; Switzerland; Greenland; and South Africa. That didn't include the upcoming expeditions to Chile and Argentina, and Admiral Byrd's next Antarctica trip. Compton's own wanderings might be completed for the time being, but his world survey was far from over.

On a more personal note, he was now finally reunited with his entire family. Back in August as they were all returning from Mexico, Betty and Arthur Alan had switched trains for Chicago in Kansas City while Compton had continued northward to Canada.

While Compton was settling back into his Chicago routine, the lengthy University of Chicago press release provided plenty of fodder for the papers to chew on for the moment. Amid the ongoing election coverage, Will Rogers witticisms, and items about the marital tribulations of singing idol Rudy Vallee and his second wife Fay Webb, readers learned that cosmic rays are the "vagrants of space," attracted by and whipping around the "giant magnet" of Earth from their mysterious, unknown origins in interstellar space or possibly also the upper atmosphere.

But most importantly, Compton had demonstrated conclusively that cosmic rays were not any sort of waves or photons, but instead some type of particles, and probably charged particles rather than James Chadwick's recently discovered neutron.[13]

Not that everyone agreed, of course, such as the editors of the *Los Angeles Times*, who deemed Compton's challenges to Robert Millikan's conclusions "both surprising and upsetting." Although they admitted that "Dr. Compton's announcement has an important bearing on the Millikan-Jeans controversy . . . the matter is far from settled." Most Caltech physicists apparently "suspected that Dr. Compton has jumped at conclusions." Anyway, how could Compton possibly make such statements, running all around the world in such a short time trying to "compile data received by wire or wireless from such widely separated points," when Millikan's theory was "laboriously developed and confirmed by year-after-year tests in widely separated regions."[14]

No matter. In case anyone still doubted the preeminence of Millikan on cosmic rays, he had now rigged up one of his supersensitive electroscopes in his own basement at home. "And now," said the *Los Angeles Times*, "we may expect soon to have the low-down on the cosmic ray."[15]

At least a basement cosmic ray outpost would be a more comfortable arrangement for Millikan. A Caltech colleague noted that on many late nights, Millikan had been "a well-known sight under a certain live-oak tree west of the Norman Bridge Laboratory, reading his cosmic-ray ionization electroscope."[16]

Before he could spend too much time tinkering in his basement, however, Millikan had to head eastward to New York to collect yet another accolade, this one a gold medal from the Roosevelt Memorial Association for being "a prophet of the new time, bearing to bewildered men, alike from atom and from star, news of the presence and the goodness of God."[17] Along the way, the "prophet of the new time" would stop off in Chicago to open the Town Hall Saturday morning lecture series at the Drake Hotel on October 22. He would speak on "New Results in the Cosmic Ray." It would be something to look forward to, because Millikan was apparently "delightful, very genial and not at all wooly-minded like some scientific men," observed the *Chicago Tribune* announcement. "Nice voice, too."[18]

Before that happened, however, Compton would give the first presentation of his adventures to a nonacademic audience, "Cosmic Rays on Six Continents," on October 19 at the University of Chicago's Mandel Hall. He had a good opening. "I am frequently asked, 'What is a cosmic ray?' I always give the same answer—'I don't know.' I am also asked, 'Where do cosmic rays come from?' and again I answer, 'I don't know.'" But that didn't mean there was nothing to talk about, and everyone could simply go home. He would instead talk about the effects of cosmic rays, "for it is by observation of their effects that we get a knowledge of their nature."

Compton proceeded to give a brief history of cosmic ray discoveries and theories, and then a travelogue of his journeys over the preceding six months, accompanied by photographs and even some motion picture footage. He would go on to give the same presentation to several other audiences in the coming weeks, including at Chicago's Union League Club, all to enthusiastic responses.[19]

It is not known whether Compton or Millikan attended each other's Chicago lectures or whether they even had time to meet up during Millikan's brief stopover, but the sage of Pasadena apparently had no lack of fans in the Windy City. On the eve of his lecture, the *Chicago Tribune*'s Ruth De Young wrote glowingly of the "friendly, lovable Dr. Robert A. Millikan." With still another mistaken attribution to him for the discovery of cosmic rays, she gushed that "his many friends here are anticipating his coming because of his own genial personality." De Young quoted an unnamed "admirer" who knew Millikan from his earlier years in Chicago: "'It is his disarming simplicity. This is where his essential greatness lies. He is a wholesome, lovable, winning person who enjoys nothing quite as much as reading Christopher Morley's "Where the Blue Begins" aloud after dinner with his family.'" The piece concluded with another encomium to Millikan's religiosity. The article would have been right at home in the pages of the *Los Angeles Times*.[20]

There would be still more Millikan enthusiasm in a *Chicago Tribune* editorial after his departure, although the writer inexplicably misnamed the scientist as "Dr. David Millikan." Still, the author admired the way that Millikan "almost played with his subject, tossing it lightly about" while looking "intensely happy, vital."[21] Millikan certainly knew how to work an audience.

One reason for his happy attitude was some recent good news from California. Another of the electroscopes sent up by Bowen from Texas the previous month had been recovered intact about seventy-five miles from its launch site and returned safely to Pasadena.[22] And Victor Neher successfully completed several more flights in a California National Guard biplane to altitudes of up to three miles, keeping an electroscope hovering over Los Angeles for more readings. Then Neher and his wife boarded the Grace Line steamer *Santa Elisa* for Peru.[23] Millikan's professional hopes and perhaps the fate of his cherished birth cry theories sailed with them.

Accepting his latest dose of adulation courtesy of the Roosevelt Memorial Association in New York (founded to honor not the current Democratic presidential candidate but rather his fifth cousin and former president Theodore), Millikan managed to combine scientific, religious, and political commentary all at the same time.

A longtime Republican and supporter of President Herbert Hoover, then fighting for reelection, Millikan noted that although many criticized Hoover's administration for appointing "so many commissions," such was something to be commended, comparable to the lessons that physicists had learned in recent decades against "cocksure thinking" and in favor of "being guided by facts and not by preconceived theories." Recent happenings in Washington, D.C., according to Millikan, were "the first time in the history of politics that we have begun to collect facts before drawing conclusions." Putting together expert commissions to examine important questions was simply a way of applying the scientific method to politics and sociology. And, said Millikan, "If we cannot do that we cannot exist as a republic." Instead, we would have to "go back four thousand years to some kind of autocracy and let somebody run us because we have not got brains enough to run it ourselves."[24]

And science had another role to play, as demonstrated by the radio technology that Hoover had used to broadcast a recent speech. It made possible "discussion on a high plane of the problems that face American life, so as to get a vote that is based on intelligence instead of emotions and prejudice."

Apparently, the Roosevelt Memorial Association agreed, presenting Millikan with a three-inch-diameter gold medal with Theodore Roosevelt on one side and flaming sword with the motto "If I must choose between righteousness and peace, I choose righteousness" on the other. The award announcement noted his Nobel Prize work and declared that Millikan's "discovery of the cosmic rays" and related work "has been an achievement equally brilliant." He had helped to move science away from "the materialistic and mechanistic theories of the nineteenth century toward spiritual conceptions in harmony with the aspirations of religion." Playing the mystic, Millikan remarked to the audience that "there is something going on in outer space which is sending messages to us that anybody can verify." Messages of an eternal universe and continuous creation, presumably.[25]

The *Los Angeles Times* editors agreed: "This doubtless was an allusion to the cosmic rays." Speaking of cosmic rays, Millikan also told reporters that he welcomed attacks upon his cosmic ray theories because such attacks were valuable in "speeding the progress of science."[26] But when pressed for more specific comment, Millikan "smilingly declined tonight to discuss reports that he and Dr. Compton had met recently and resolved their differences."[27]

Whatever the source of such naive and forlornly hopeful reports, Millikan could also rest assured that along with his gold medal and all the other positive news he had received in the past week or so, he was still also considered the number one cosmic ray champion in the *Los Angeles Times* sports pages, whose reporters were confident that in their upcoming contest against their undefeated rivals the Redlands Bulldogs, the Caltech Engineers football team would "blossom out with an attack that will feature forward and lateral passes along with plays said to involve knowledge of the stratosphere and cosmic rays garnered from Prof. Millikan."[28]

Arthur Compton's name was rarely if ever seen in the Chicago sports pages, nor was the scientist reportedly at all concerned with lateral passes and football rivalries, despite his university's role as a founding member of the Big Ten college athletic conference.

Instead, when he wasn't busily collecting, collating, and analyzing the data from his own journeys and the others in progress or giving his "Cosmic Rays on Six Continents" presentation to rapt audiences, Compton was getting ready to report to his most important audience of all: his fellow scientists. The upcoming annual meeting of the National Academy of Sciences (NAS) was coming up in mid-November, and it was at this event that Compton had chosen to formally go on the record to the scientific community. He would make several definitive pronouncements, one of them wholly correct, one only partially true, and the other flat wrong, although he wouldn't realize it until sometime later.

If they hadn't already gotten the message, it was now clear even to the members of the press who didn't specialize in science that a major confrontation was brewing and was soon to boil over. Perhaps as a sort of journalistic scene setting, and to provide some helpful up-to-date background for those who hadn't been following the whole cosmic ray saga very closely, *New York Times* science editor Waldemar Kaempffert provided an extensive summary on the subject in a Sunday supplement piece right before the NAS meeting was to begin. He opened with a quote from Dr. Thomas H. Johnson of Philadelphia's Bartol Research Foundation of the Franklin Institute in Philadelphia, another center of cosmic ray research that had suffered somewhat in the shadow of Millikan and Compton's institutions. "Never before in the history of science has there been a subject about which so many able investigators have disagreed," Johnson said.

From that apt and accurate beginning, Kaempffert ably laid out a brief history of cosmic ray history, research, and the current controversies, accompanied with a lucid explanation of the science involved. The long feature was similar in intent to William Laurence's piece on another Sunday two years previously, although Kaempffert's take was considerably more balanced and far less Millikan-centric than Laurence's article. Then again, it was certainly true that Millikan was the dominant figure in the field at the time Laurence was writing. Kaempffert's piece, perhaps unintentionally, makes it clear that only two years later, Millikan's status as top gun in cosmic rays was under serious challenge from Compton, Jeans, Regener, and other notables.[29]

The next day, Compton spoke before the NAS meeting in Ann Arbor, Michigan. He described the work and conclusions of Millikan, Jeans, and others, and then the results of his world survey. Not only were cosmic rays charged particles, likely electrons, there was definitely a latitude effect, with more rays detected in the polar rather than the equatorial regions, findings wholly inconsistent with the photon hypothesis. Compton also announced that the preponderance of the evidence seemed to indicate that cosmic rays likely originated not from deep space, but from the Earth's own atmosphere, probably about a hundred miles up. Still, he allowed that some percentage of the most energetic rays might be coming from deep space as well.

"It would thus appear that the data which have been obtained in this geographical study of cosmic rays may be satisfactorily explained if we suppose that the cosmic rays consist of electrons originating some hundreds of miles above the surface of the earth in the upper stratosphere," said Compton. "It seems very difficult to reconcile with our data any of the alternative hypotheses that have been suggested."

He was not, however, claiming to have all the answers about cosmic rays. "What their message may be, we have not learned," Compton noted. "Perhaps they are telling us how the world has evolved or perhaps they are bringing news of the innermost structure of the atomic nucleus."[30]

Suddenly, it seemed, cosmic rays were not so cosmic, not harbingers of universal creation and an eternal universe, and definitely not Millikan's birth cries of atoms from somewhere out in the reaches of infinity. If what Compton was saying about their origins was true, then they were far more prosaic, as mundane and commonplace as the aurora borealis or a summer thunderstorm. So much for all the talk about God and eternal truths and science showing the way to religious verities.

Compton had perhaps jumped the gun somewhat. For one thing, his world survey was not yet complete; more data was yet to come from the remaining expeditions over the coming months. For another, he was perfectly aware that other colleagues, not all of them named Millikan, had other reasonable hypotheses and some interesting data to back them up. To his credit, aside from his confidence in the latitude effect, his NAS presentation was liberally sprinkled with provisional qualifying remarks such as "the preponderance of evidence" and "it would thus appear" and "it might well be." He was not making categorical statements of eternal truths for the ages; he was clearly open to changing his mind upon new and better evidence. He would do just that in the near future.

But for now, he knew that he had decidedly thrown down the gauntlet to Millikan. In the meantime, however, the Greek chorus of the press would have their say. The editors of the *New York Times* were apparently impressed at Compton's professionalism and work ethic. The *Los Angeles Times*, predictably, was not.

"One thing not to forget while listening to Professor Compton discuss cosmic rays is that modern physics is not all speculation. A tremendous amount of leg work and eye work is involved," said the *New York Times*. "The general public is not altogether to be blamed for smiling now and then at the ease with which the physicists have been pulling strange new universes out of very small silk hats. . . . It is decidedly a come-down, after being invited to contemplate the wondrous glory of the new creations of the new physics, to be reminded by the creator himself that of course we cannot say with confidence that we really know what we are talking about."

But that was to be commended rather than condemned, the *New York Times* said. Even if Einstein seemed to build whole universes out of nothing more than a pencil and piece of paper, any errors he might make "would soon be run down by the boys with the surveyor's rod and the telescope and the x-ray camera. . . . Our scientists go in for a lot of speculation, but they also do a lot of hard work. Unlike speculators in other fields, they don't seem to mind in the least being deflated by a hard fact. They seem to love it. A hypothesis slain by a fact is not only Tragedy, it is also Science."[31]

Unless, perhaps, all of a scientist's hard work was directed toward attempting to deflate someone like Robert Millikan. Observing that "certain European scientists" had been attempting to disprove Millikan's cosmic ray theories, the *Los Angeles Times* noted that "nobody has gone to quite so much trouble as Dr. Arthur H. Compton," who was getting "front-page reports of his attack upon the Millikan hypotheses." The "learned professor" from Chicago offered no proof of his "rather vague postulates," just "another hypothesis." He was, in effect, only guessing when he spoke of electrons in the upper atmosphere. "The studious mind, therefore, is likely to shy at the Compton guess and be more ready to accept the Millikan theory as to the cosmic origin of the rays, which, so far as has been actually demonstrated, remains untouched."[32]

While the editorial pages seemed to be preparing to take sides in the impending scientific prize fight of the century, Millikan received some big news. Now he was ready. The Chief in Pasadena was about to pick up Compton's gauntlet.

11

THE GATHERING FORCES

One of those "certain European scientists" mentioned by the *Los Angeles Times* who had been daring to challenge the theories of Dr. Millikan remained the ebullient Auguste Piccard, who true to his announcement following his record-breaking stratospheric voyage over the summer, was proceeding apace with his plans for a more northerly excursion over Canada's Hudson Bay. Although that earlier flight had convinced him that cosmic rays originated in the upper atmosphere, he still wanted to collect further data at the polar regions.

Not personally, however. He was now seeking "two husky Americans accustomed to camping in the coldest weather" to man his balloon for the Hudson Bay flight, along with an American millionaire willing to finance the venture. Preparing to head to the United States for a lecture tour and visit to his brother Jean in Delaware, he announced, "All I need for the Hudson Bay flight is a new, perfected balloon, money, fine bright weather with no fog, and two men even much more fitted than I am to stand the terrific cold above the magnetic pole." He hoped that the men would be both pilots and physicists.[1]

Robert Millikan and Arthur Compton were unquestionably both physicists, but neither were licensed pilots, despite Compton's youthful enthusiasm for building airplanes. And while they had both displayed a zest for travel and adventure, they weren't about to go personally probing the stratosphere in a balloon. They already had a previous engagement in Atlantic City for the annual meeting of the American Association for the Advancement of Science.

Before that happened, Robert Millikan would finally break his almost complete silence of the past six months since Compton published the first results of his world survey in *Physical Review*. Not that Millikan was ever lacking in general self-confidence, but now he had received welcome news that specifically boosted his confidence about the correctness of his cosmic ray theories and the wrongness of Compton's.

On November 10, Victor Neher's steamship reached Balboa, Panama, at the Pacific entrance to the Panama Canal. Neher had been under Millikan's instructions to take electroscope readings all the way down from Los Angeles and then report back as soon as he reached Panama. But telegrams from distant lands were expensive, and Millikan had specified that he wanted only the headlines, nothing more. When every single word cost money, the details could wait until later. As Neher put it, "Millikan was not one to depart from the reputation of his Scottish ancestry, and he had instructed me to be as brief as possible."[2]

Unfortunately, Neher found himself in a situation difficult to report in only a few words. Shortly after departing Los Angeles, when Neher was attempting to take readings in the same region where Millikan had failed to collect data on his 1926 voyage to South America, the quartz system in the electroscope broke, a casualty of the rush to get everything ready before departure. "This was a horrible thing to happen," recalled Neher. "Millikan, six years earlier, had missed this strategic region, and now we had done the same."[3]

Neher had a spare, but the replacement procedure was too delicate to be performed on a rolling, pitching ship, so he had to wait until the ship anchored briefly at the bay in Mazatlán, Mexico. A day after departure, he began taking readings again. Everything seemed to be working fine, although he still had to make some adjustments.

So far, all the data Neher had managed to collect from Los Angeles to Panama had shown no evidence of a latitude effect. But he knew that those results were considerably compromised by the technical mishap with the electroscope that had blanked out a large part of the data collection on the southward voyage. How could he explain to the Chief all that had happened in a brief cable?

Neher knew the main thing that Millikan wanted to hear about was the latitude effect. He decided to send a two-word telegram just so Millikan would get the main point and send a follow-up letter explaining everything else afterward. Neher's November 10 cable from Balboa was a model of succinctness: "No change." In other words, no change in electroscope readings with the change in latitude from Los Angeles to Panama. Ten days later, when Neher reached Mollendo, Peru having made further measurements along the way, he sent Millikan an identically worded telegram.[4]

That was all Millikan needed to speak out at last. He chose his home ground, a talk before the Astronomy and Physics Club in Pasadena on December 2 at the library of the Mt. Wilson Observatory, perhaps partly because he knew he would have a sympathetic audience there, without facing the contention that was sure to come soon in Atlantic City. The *New York Times* reported that Millikan "answered his rival's challenge," which had "threatened to demolish his own research results."[5] He again declared that cosmic rays were photons, not electrons, and that any electrons or other particles that happened to be detected in electroscopes weren't actually cosmic rays but merely particles knocked out of atoms by the rays. As to a latitude effect, said Millikan, "all the most careful experimenters have attested to the fact that there is no detectable influence by the earth's magnetic field," at least one that his measurements had detected, although he allowed that perhaps "a certain small variation with latitude is to be expected at the earth's surface under certain conditions." There was also "no direct evidence on the cosmic ray intensities" exerted by the sun.

Millikan refrained from directly naming Compton or Jeans, the colleagues trying to "demolish" his results, but at least according to the *Los Angeles Times*, "offered such startling new proof of the correctness of his previous work as to cause the distinguished audience to approve his answer with spontaneous applause," an audience composed of "virtually every one of this city's 100 noted astronomers and physicists."[6]

As usual, the *Los Angeles Times* played cheerleader for Millikan on its editorial page several days later. Although Arthur Compton had been leading "a little group of American scientists" who had been busy trying to upset Millikan's theory and had "published thesis after thesis and lectured before several scientific bodies in support of his own theory and in contravention of that of Dr. Millikan," the Pasadena scientist had now demonstrated that he was correct and that Compton's electrons were only the atomic detritus of photonic cosmic rays. Any latitude effect was "so small as to be negligible." And furthermore, the *Los Angeles Times* noted, Millikan's evidence that the sun has no effect on cosmic rays disposed of the arguments of the distinguished Sir James Jeans as well. All was well with the Pasadena group of scientists, whose evidence "doubtless will be accepted by physicists generally."[7]

After laying low for a while, Millikan was back in the savant spotlight. "Thus he regained the attention of the press," wrote De Maria and Russo. "The public took a new interest in his controversy with Compton."[8] In only a few weeks, that interest would become a storm that would persist well into the coming new year, and Millikan would soon have cause to regret jumping the publicity gun before getting all the details of Neher's southerly adventures.

As interested and involved parties in the realms of science, press, and public settled in to take sides, others were already striving to facilitate some sort of reconciliation. Most prominent was the thirty-eight-year-old Belgian mathematician, astronomer, physicist, and Jesuit priest Georges Lemaitre, perhaps the most unusual of the scientists then pondering the biggest questions of the universe, a group that included Einstein, Edwin Hubble, Arthur Eddington, James Jeans, and of course Robert Millikan. Lemaitre laid much of the mathematical and theoretical framework for the expanding universe that Hubble had discovered and would eventually be credited as the originator of the Big Bang theory.

In December, Lemaitre was visiting Pasadena and Caltech, meeting Millikan in person for the first time. Having already met with Compton in Chicago and discussed the recent cosmic ray questions, he was sure that it was possible to reach some sort of scientific *modus vivendi* between the two Nobel Prize winners. Although he agreed with Millikan that cosmic rays came from outer space and weren't products of the upper atmosphere as Compton had recently postulated, Lemaitre also held Compton's view that they weren't photons but instead some variety of charged particles.

After lecturing to the Pasadena crowd at Caltech and Mt. Wilson Observatory on his concept of a primordial atom giving birth to the universe, he lunched with Millikan at Caltech's Atheneum Club, where the bill of fare was "alpha-ray soup, cosmic salad and luscious nebulae frappe," according to a reporter. Some speculated that Lemaitre was actively trying to play the role of peacemaker between Millikan and Compton by showing ways to reconcile their opposing positions. After two hours over lunch with Millikan, however, Lemaitre reported, "I have had a most interesting discussion, but there is nothing special to report."[9]

Despite the best efforts of the scholarly Belgian priest, however, neither Millikan nor Compton seemed much inclined to be moved. At the 181st meeting of the American Physical Society in Pasadena a little over a week later, Millikan continued to make his case, aided by further calculations by Caltech's R. M. Langer.[10] Meanwhile, Ralph Bennett was presenting the Compton case to the public at popular science lectures in Cambridge, assisted by "John Cosmic Ray," who passed through various instruments and camera and rang a bell for the audience's amusement.[11] Later in the month, Bennett continued his observations for Compton, leading a party of scientists in an all-night cosmic ray observing vigil atop one of MIT's buildings, part of a network of fifty identical overnight vigils.[12]

On Christmas Day, *Los Angeles Times* science columnist Ransome Sutton took note that the papers presented at the recent National Academy of Sciences meeting the month before were starting to appear, including Compton's report on his global survey. "But his paper probably does not express the 'last word' on this controverted subject," Sutton remarked mildly.[13]

It was definitely not the last word on the subject. Robert Millikan and Arthur Compton were about to meet up in the seaside resort town of Atlantic City, New Jersey, to decide before an audience of colleagues, press, and the public just who was going to get the last word on cosmic rays.

In December 1932, Atlantic City was not the East Coast gambling haven it would become later in the twentieth century. Nor was it at all known as a center of science and learning. Instead, it was famous for tourism, which included nightclubs, liquor even during the waning years of Prohibition, and sometimes general debauchery. It was also known for the world's first boardwalk, its seaside climate, saltwater taffy, and its beaches, shops, and resorts. As a 1922 tourist brochure dubbed it, Atlantic City was "the world's playground."

In short, it was not the sort of place that one would expect to host a major professional scientific conclave. On the other hand, the city was easily accessible by rail, car, and air from major metropolitan areas such as New York, Philadelphia, Baltimore, Boston, and Washington, D.C., making it a ready destination for attendees coming from all the major academic institutions in those areas or making connections from elsewhere. And the town certainly had no lack of hotel rooms, meeting spaces, or entertainment facilities for those so inclined in their off hours. The AAAS held its annual meetings in different cities each year to give scientists in all parts of the country a chance to attend, meet each other, and take in the sights, both scientific and touristy, of each unique location.

So Atlantic City was a reasonable choice for the thirty-ninth AAAS annual meeting. From December 27 to December 31, 1932, there would be lectures, symposiums, exhibitions, experiment demonstrations, and parties and informal gatherings, attended by prominent scientists from the United States and beyond. Forty-one of America's other major scientific societies such as the American Physical Society, American Astronomical Society, and Mathematical Association of America would hold meetings there as well. Scientific publishers and commercial companies would exhibit and present their wares. There would even be day-long field trips to nearby scientific centers such as Princeton University.[14]

As all the scientists, academics, students, job-seekers, reporters, and other interested parties began to descend upon what was then called "The World's Playground" in the days after Christmas and mix in with the crowds of tourists, gamblers, showgirls, hucksters, gangsters, and con artists, however, the most anticipated event of the coming week was the cosmic ray symposium on Friday. Robert Millikan and Arthur Compton would go head to head live and in person, each preaching his own cosmic ray gospel to the congregation of fellow scientists, facing each other directly instead of by proxy through competing press coverage.

And the papers were loving it. Not that the press or public really cared all that much whether cosmic rays were photons, electrons, or something else entirely, or about birth cries of elements versus the death wails of stars, or about magnetic poles and latitude effects. The fascinating

11.1 The rivals: Robert Millikan (left) and Arthur Compton, 1932. (University of Chicago Photographic Archive, [apf1–01870], Hanna Holborn Gray Special Collections Research Center, University of Chicago Library)

phenomenon here was the spectacle of two eminent Nobel Prize winners locked in an increasingly intense and bitter debate just like a couple of local politicos arguing about zoning laws—and the realization that by definition, one of them had to be wrong. And that meant that after all, scientists were not infallible geniuses who knew everything. Not only could they be wrong, they also could be petty, stubborn, self-serving, and make bone-headed mistakes just like the rest of us puny mortals.

According to literary theorists, conflict is the essence of drama. It is also a highly effective way to sell newspapers. But political battles between presidential candidates or violent confrontations between cops and gangsters were fairly straightforward: Democrat vs. Republican, good vs. evil, and so on. How to portray this clash of titans between two of the most brilliant and celebrated scientists in the world? Was it a heavyweight title fight? A learned debate between scholars? A theological disagreement over matters of doctrine? A knife fight between two hoods in a grimy alley? All or none of the above?

The *New York Times* went the "learned debate" route with reasonably subdued coverage, though they noted in a subheadline, "Clash of Millikan and Compton to Form High Point at Scientific Convention."[15] Other outlets tried to have it multiple ways. The *Boston Globe*, for example, headlined its Associated Press coverage of the upcoming meeting with "Debate is Ahead on Cosmic Rays," as if it were nothing but a mild student council debate, but added the subheadline, "Compton and Millikan Will Clash Friday," which gave the event something of a prize-fight sheen. But the article itself cast the confrontation in yet another light: "Before scientific America's Supreme Court will go this week a question debated in councils of the learned and before firesides of the laymen—the what and whence of cosmic rays. The tribunal is the annual meeting of the American Association for the Advancement of Science in Atlantic City."

As Compton "girded himself for the clash," he apparently took time to speak to the Associated Press and lay out the basic parameters of the issues at hand and "discussed the significance of the debate." He emphasized that despite appearances, there was really no actual conflict, nor would any questions be definitively resolved. The chief goal of the symposium was to stimulate discussion and further research. As to why anyone should care at all about cosmic rays, since they had no immediate meaning or application to the lives of most people, Compton explained that "the very sound of the name suggests something huge and vital—cosmic in importance. To some it has a spiritual connotation." Also, "people want to learn about something that has been with them since the beginning of time but which only in recent years has entered the knowledge of mankind. It's just curiosity."[16]

The *Globe*'s editors seemed to applaud the spiritual connotations, at least. "Learned societies meet in the ball rooms of excessively comfortable but otherwise not much frequented beach hotels and tell each other what they know," said an editorial. "Four thousand, five hundred scientific scholars are gathered in Atlantic City today for such an exchange, and the great event of it is to be a debate between two eminent scientists on the nature of the cosmic rays. These meetings happily convert columns of newsprint into a species of university extension course."

Did all that endanger religious faith? Not at all, said the editors, relating a tale of an old gentleman who had lost his faith in traditional theology yet remained more religious than ever. "And what, if one may ask, is the source of his religious consolation? Cosmic rays."[17]

Another briefer editorial comment noted waggishly, "Not much seems to be known about either cosmic rays or amnesia, except that the latter is of more practical use."[18]

Millikan was a long-experienced hand at dealing with the press, and usually able to steer things the way he liked them, but he was also perceptive and shrewd enough to see what was coming, even from a great distance. That was likely one reason he had mostly avoided the press for months following the conclusion of his cosmic ray expeditions earlier in the year, letting Compton have his way in the spotlight while he hunkered down.

And, as press and public interest continued to intensify over the past weeks and months and to focus more on the Atlantic City encounter, it was certainly the reason why he reached out to Compton a month earlier, attempting to ward off or at least soften the impact of the inevitable.

As soon as he received Neher's second cable from Peru on November 30, he sat down and wrote to Compton, telling him of the reports of "no change" in cosmic ray intensity even down to the equator. With studied politeness, he suggested to Compton that "in view of [these results] you may want to at least state some of your own results somewhat differently than you otherwise do." He proposed that the two of them avoid any discussion of the latitude effect at Atlantic City "until some later time when the avidity of the newspapers has died down and we can get the records into the scientific journals without attracting the attention of the newspapers at all." After all, Millikan noted, he had already pointed out more than a year ago that some small latitude effect was to be expected anyway and wasn't that important. Considering that, "there is nothing left for controversy" and in Atlantic City they could "confine attention to facts upon which there is general agreement." Instead, they should simply "tell the technic [sic] which we are using and withhold the results."

Millikan was still trying to find some way to reconcile a latitude effect with his cherished photon hypothesis, and thought it was yet possible. "I can admit the possibility of some equatorial latitude effects so that we can appear before the public as not having got contradictory experimental findings," he wrote. "In a few months the newspapers will have lost their interest and the wholly exaggerated and unfortunate emphasis upon controversial results will have disappeared."

Millikan closed with a wish that Compton would agree on the wisdom of such a course and that the "advance knowledge" he was sharing would help Compton in "developing a form of presentation of your own results which may not come back to plague anybody later."[19]

Compton was having none of it. On December 5, three days after Millikan had publicly "answered his rival's challenge" in Pasadena, he responded with a respectful but firm demurral. Thanking Millikan for the advance results, he also reminded his colleague that he had just completed six independent comparisons of cosmic ray intensity in both equatorial and temperate zones, and they all consistently showed a definite latitude effect. While he also regretted the "exaggerated emphasis" that the newspapers were placing on the whole issue, he didn't see how it could be kept under wraps at Atlantic City. "In fact, I can not see how it is possible at the present time to keep from the newspapers the fact that evidence is on hand indicating marked differences in intensity at different latitudes, and that this points toward an electrical particle nature for the cosmic rays." He mentioned that the fact that his initial letter to the *Physical Review* back in July had made its way into the papers "suggests how difficult it would be to keep such information from the news hounds." And even if he and Millikan were to hold back on announcing their findings, other researchers in Europe were investigating the latitude effect, and it was only a question of where the evidence would be published first, Europe or the United States.

Compton, therefore, was not prepared to soft pedal or downplay anything. "I do not consider our own experimental data to be provisional," he stated. His results "have been checked and rechecked, using all the precautions I have known how to take. They have always given the same result." The intense scientific interest in the issue made it unwise and impossible to suppress any data from the public. The Atlantic City symposium would and should help to reveal any flaws in anyone's data, along with making it possible to make any necessary adjustments or come up with any possible alternative interpretations. "I feel that any other procedure would appear so much like dodging the issue that we should both be blamed for lack of scientific candor."[20]

Compton's response may also have been influenced somewhat by the fact that by the time he answered Millikan's original entreaty, Millikan had already come out swinging by announcing his own results and affirming his own position to the press. In any case, it didn't matter. No matter what optimists such as Lemaitre may have hoped, a clash in Atlantic City could not be avoided.

The cosmic ray symposium had actually been scheduled for Wednesday, December 28, 1932, at 10:00 a.m., but for some reason, possibly the expected large attendance, got rescheduled to Friday. No matter; it gave both Compton and Millikan a couple more days to "gird their loins," prepare their charts and data, and contemplate the meeting.

Some of the attending members of the Mathematical Association of America may have been disappointed at the change of schedule, since they had a field trip to Princeton University already set for Friday, departing from Atlantic City's Pennsylvania Station at 8:45 a.m. There they would attend their own symposium on calculus and mechanics, inspect the new mathematical building, and attend a lecture by genius mathematician John von Neumann.[21] They would have to hear about the Millikan-Compton debate after the fact.

Finally, it was time for the showdown to begin. William Laurence, of course, was there for the *New York Times*, eagerly anticipating the fireworks. "In an atmosphere surcharged with drama, in which the human element was by no means lacking, the two protagonists presented their views with the vehemence and fervor of those theoretical debates of bygone days when learned men clashed over the number of angels that could dance on the point of a needle," he would soon write.[22]

Millikan's confidence as he proceeded to enter the ring had been profoundly shaken, however. As previously arranged, Victor Neher had completed his readings in South America, on the ground, in the mountains, and flying at 19,000 feet on a Ford Trimotor of Pan American-Grace Airways. Then he and his wife boarded a ship at Mollendo and proceeded north to New York City through the Panama Canal, taking further measurements all the way.

"Going through the Caribbean and on up the east coast of the United States it became clear that there was a latitude effect," he recalled. On

December 27, at 43 degrees north, he dispatched another cable to Millikan, who was by then in Atlantic City: "Seven percent change returning. Concealed before by broken system and different ships." He was referring both to the quartz system problems on the way south from Los Angeles and to the fact that, as he explained, "in changing ships there is always the uncertainty of different absorbing materials in the walls and overhead. There is sufficient difference so that readings cannot be readily compared."[23]

He wasn't looking forward to facing the Chief when he finally disembarked in New York and headed down to the AAAS meeting in Atlantic City. On arrival, he bumped into Caltech colleague Richard Tolman, who asked him whether he'd found a latitude effect. Yes, said Neher. Tolman clucked his tongue and shook his head, knowing all too well how Millikan would take the news.[24]

It is unclear whether Millikan actually received Neher's telegram or saw him personally in Atlantic City before or after the debate with Compton. But whatever the case, Millikan was shaken up enough to change Neher's travel plans back to California after the meeting. "The original plan was to return by train to Pasadena," said Neher. Now, however, Millikan wanted more data to make up for the problems Neher experienced on the outbound voyage. Instead of heading straight home by train, "it was decided to take a ship back through the canal. Arrangements were also made with the Army Air Corps to make an airplane flight at Panama."

It wouldn't make any difference, as Neher later recalled. "The data from New York through the canal and up to Los Angeles were consistent with the data taken from Mollendo to New York. Also, the results of the airplane flight at Panama were consistent with the flights made in Peru."[25]

For the moment, though, that was in the future. Millikan first had to face the man who had become his scientific arch-nemesis with a considerable disadvantage. Fortunately, his rival didn't yet know about it.

12

HIGH NOON IN ATLANTIC CITY

Compton went first, with his innocuously titled presentation of "Some Evidence Regarding the Nature of Cosmic Rays." After reviewing the previous work that had been done in measuring cosmic rays across widely separated geographical regions, including by Dutch researcher Jacob Clay, Millikan and Cameron, and Bothe and Kolhörster, he noted that those experiments had found little if any differences, although Clay had claimed to see a latitude effect. Given the strong evidence from the Bothe-Kolhörster coincidence counter experiments that cosmic rays were some kind of charged particles, however, the failure to confirm any sort of geographical variations "was of unusual interest."

In the published version of his presentation that later appeared in *Physical Review*, Compton noted that if those negative results could be pinned down with greater precision, "it would mean that the cosmic rays could not be electrical particles coming from outside the earth's atmosphere"—which was, in part, what Millikan had been claiming all along. On the other hand, if a latitude effect such as Clay had claimed did exist, it would support "the assumption of high-speed electrical particles entering the earth's atmosphere."[1]

In an effort to resolve the anomalies, Compton and several colleagues had organized "a group of expeditions that have attempted to make as extensive a study of the geographic distribution of cosmic rays as could be done in a limited amount of time. Eighty-one stations at which measurements have been made are about equally divided between the Northern and Southern Hemispheres. They have extended from latitude 46 degrees south to latitude 78 degrees north and from sea level to about 20,000 feet." Compton displayed a world map indicating all the measurement locations and described in detail the instruments and experimental techniques used.

All the collected data taken together, Compton announced, "show a marked difference in intensity between the cosmic rays at temperate and polar latitudes, as compared with the tropical latitudes." That difference was also evident at different latitudes and actually was more marked at high altitude between the polar and equatorial regions.

The experimental findings seemed to be consistent with the theories of Lemaitre and Vallarta that the cosmic rays falling upon the earth could be placed in two groups, one of very high energy and penetrating power and the other less penetrating but still quite powerful at about seven billion electron volts; the former could be photonic while the latter were likely electrons. Either way, the energies Compton was discussing were far greater than those Millikan claimed were possible (and which, not incidentally, supported his birth cries theories).

"This geographic study of the cosmic rays thus indicates that the less penetrating part of the cosmic rays, at least, consists of high-speed electrified particles," said Compton. Such a conclusion was supported by the ionization curves he had observed at different altitudes as well as the high-altitude measurements of Regener's sounding balloon flights. Although the current theory seemed to indicate that these "electrified particles" were electrons, Compton allowed that "the experiments that have been cited do not serve to distinguish between negatively and positively charged particles," but were equally consistent with protons or alpha particles.

But, he concluded, "I find no way of reconciling the data, however, with the hypothesis that any considerable portion of the cosmic rays consists of photons."

By all accounts, Compton's presentation was calm, measured, matter of fact, and devoid of any overt personal commentary, editorializing, or gratuitous sideswipes at his scientific opponent. If it was true that, as one physicist had put it before the symposium, "Millikan has a chip on his shoulder and Compton is . . . [ready] to knock it off"[2]—and no doubt it certainly *was* true—Compton was not going to let it show and instead relied solely on the strength of his data and research to make his case.

Robert Millikan, however, did indeed have a considerable chip on his shoulder, and he was not prepared to let anyone knock it off, least of all this junior colleague and relative cosmic-ray neophyte from Chicago—Nobel laureate or not.

The title of Millikan's presentation was anodyne enough, as objective and scientifically neutral as Compton's: "New Techniques in the Cosmic-Ray Field and Some of the Results Obtained with Them." If anyone in the audience found that disappointing and feared that what was to follow would be similarly dull and boring, without any of the fireworks everyone had been anticipating, they were soon reassured. Millikan came out swinging, yet with a decided defensiveness, an aggrieved authority who was in equal parts bewildered, mystified, and annoyed by the parties arrayed against his previously inviolate authority.

He began with a long and somewhat belabored analogy comparing the development of a new scientific field to a "man child." At first, he said, everyone is intensely interested in the boy because he is new and charming, an awakening presence slowly revealing the person he will eventually become. Then by about age eight he becomes a whining brat for a while, until by about age thirteen he begins to settle down, until by age twenty-one he is "ready to don the toga and be admitted regularly into the ordered society of adults."[3]

The cosmic ray field was now at about that age, Millikan observed, and thus should be ready to "don the toga" as a mature discipline. But because the field had been robbed of six or seven years of development by World War I, it was instead just emerging into maturity. Until that was complete, the less attention paid to the "ugly brat," the better.

"I shall then have little to say today about the controversial aspects of the cosmic ray field," Millikan declared, which would soon be cleared up anyway, and were not "anything like as controversial now as the public, for some reason, seems to think." In other words, this whole commotion about a huge argument between scientists was simply exaggerated, because "so far as all the major and really significant facts of observation are concerned, there is general agreement now." Any disagreements were only matters of interpretation or accuracy.

As far as Millikan himself was concerned, "nothing has happened which alters in any essential way the views expressed in my last comprehensive report, written more than a year ago," and he hoped that anyone interested in the subject would read that report before expressing or forming any opinions, "for within the past ten months the American newspapers have been quite consistently misunderstanding and misrepresenting my findings, and I suspect that some of my physicist friends, who have been too busy with their own work to look up the facts, have been pretty thoroughly misled by incorrect reports." Whatever he said today, then, should be regarded simply as a supplement to his previous statements, and anything new served only to confirm and amplify what he had already reported.

Having therefore sufficiently established his status as injured party, Millikan proceeded to spell out the "important and non-controversial" facts upon which everyone obviously already agreed. Among those were that the charged particles that researchers such as Compton had been measuring and reporting as showing a latitude effect weren't actually the primary cosmic rays coming from outer space, but secondary particles created by primary rays, that is, photons, colliding with atoms in the upper atmosphere. This, he said, was exactly what had been seen and recorded in cloud chamber experiments such as those resulting in Carl Anderson's positron discovery. The measured energies in such experiments also supported such an interpretation. "If others, who have duplicated our technique, claim to measure energies higher than a billion electron-volts, Dr. Anderson and I think that this means that they are less cautious in their estimates than we are, and this, in turn, presumably means that they have had less experience than had we in studying the minute irregularities in the straightness of cosmic-ray tracks." In other words, went the implication, perhaps his opponents were simply incompetent.

Millikan discounted the evidence of Bothe and Kolhörster's coincidence experiments that cosmic rays were particles, and he had already been saying for two years that "these counter experiments never in my judgment actually measure the absorption coefficients of anything," and could therefore be dismissed as spurious and meaningless.

Regarding the much-discussed latitude effect, Millikan first noted that he and Cameron had already looked for it back in 1926 and found nothing beyond the expected margins of error. Not only that, but he had also pointed out a year ago that "new light should be thrown upon this question," and he had applied for Carnegie funds to do just that "long before any latitude controversy was ever dreamed of." He notably omitted any mention of other findings that supported the latitude effect, not only Clay's but also those that had just been presented by Compton.

Historians De Maria and Russo found that the typewritten text of Millikan's presentation included the line, "Dr. Neher has just carried one of our new sensitive recording electroscopes from Pasadena to Mollendo and back without noting at sea level any changes outside the limits of our uncertainties." That may have been what Millikan had been hoping when he originally penned that line, but of course, it was not the case. "We doubt that this sentence was uttered in Atlantic City," De Maria and Russo noted wryly.[4]

Millikan had to finesse the evidence of his own high-altitude airplane measurements from Canada and California and Neher's South American sojourn, all of which had shown evidence to support the latitude effect. He latched onto a small difference between day and night measurements as a possible indication of some kind of solar effect at high altitudes. He was reaching, but he also could not actually falsify his already-published data. "These reports on the high altitude airplane effects must then be considered as preliminary and provisional," he pronounced. "A few months hence we shall have more and better information upon this point."

Otherwise, he stated, as far as sea level observations, he had always been very careful to say only that he had never found any latitude effect outside of the observational margin of error. "That statement I shall obviously never have to retract and it cannot properly be made the subject of controversy," Millikan said hopefully. He accepted that he might have to change his estimated uncertainties at some time in the future, but "if the time ever comes in which I should wish to do so, I hope I may be generously given the privilege without being too seriously pilloried or being subjected to too public an exposure of the fallibilities of my judgment." Millikan was playing the victim once again.[5]

Notably absent from Millikan's presentation was any mention of his birth cries theories. He stuck solely to cosmic rays, both extolling and defending his own techniques, results, and conclusions, while discounting, playing down, or outright attacking those of his colleagues. William Laurence was one reporter who noticed: "In upholding the photon theory of the cosmic ray, Dr. Millikan was at the same time championing another cause, though he did not mention it today. To him the cosmic ray 'furnishes some experimental evidence that the Creator is still on the job [Laurence was here referring back to the article that got him into trouble back in 1930],' that the rays are really 'birth cries' of new matter. . . . But this hypothesis is closely bound up with the assumption that the cosmic rays are photons. If they are electrons, such a hypothesis of new creation becomes untenable."[6]

Compton and Millikan weren't the only presenters at the symposium. Thomas H. Johnson of Philadelphia's Bartol Research Foundation also presented evidence supporting the Compton position of charged particles and the latitude effect and against Millikan's photons, and several other brief presentations followed.[7]

But there was no question that the main event had been Compton vs. Millikan. Unfortunately, however, there seemed to be no clear victor, no gloved hand to raise in triumph. "Both presented a mass of detailed arguments, photographs and charts to support their views, and the result was 'no decision,'" reported the Associated Press. "The outstanding fact developed by the debate was that solution of the origin of the rays must await further investigation."[8] The *New York Times* also called it a draw in an editorial, though giving Compton the edge: "We have our choice between two perfectly self-consistent theories, with most authorities leaning latterly toward Compton's because it involves no abandonment of that law of physics which holds that the universe is subject to a process of degradation rather than of ever-new creation."[9]

It may have been a frustrating outcome for the various observers more personally invested in the battle than the science, but William Laurence, at least, was finely attuned to the nuances of the conflict. "Dr. Millikan particularly sprinkled his talk with remarks directly aimed at his protagonist's scientific acumen," he reported in the *New York Times*. "There was obvious coolness between the two men when they met after the debate was over."

That coolness was quite apparent in a press photograph apparently taken at the time of the debate, although whether before or after is uncertain. Millikan and Compton are facing off, the silver-haired Millikan in his characteristic bow tie and tweedy suit looking every bit the academic patrician, the much younger and slightly taller Compton in a darker, more stylish suit and necktie, notes held under one folded arm. Neither are smiling, but are clearly deep in animated, serious, and what appears to be only slightly cordial conversation.[10]

For whatever reason, perhaps because he knew both men and felt some misguided urge to play diplomat, Laurence chose to step directly into the fray. "I asked Millikan to shake hands with Compton for the photographer," he remembered. "Millikan refused."

So much for diplomacy. Laurence, as well as all the other reporters who had been on hand, went off to write and file their stories. "I gave a very colorful story of this great debate with two great men," Laurence said. "And then to make absolutely sure of accuracy, I gave 1500 words extra for Compton's and 1500 words of Millikan opposite each other so there could be no question at all of what they said, and I devoted my story to color. That night as I wrote the story I got a very complimentary telegram from my night managing editor."[11] The *New York Times*'s coverage was indeed thorough and balanced, extensively quoting from the presentations of both men as Laurence noted.

Unfortunately, that was not enough for Robert Millikan. The next day, New Year's Day 1933, he began the new year by dispatching a long, sharply worded telegram to the *New York Times* editors, attacking the paper's coverage of the debate in general and Laurence's article in particular.

After noting that he customarily did not try to correct erroneous newspaper reports and praising the *New York Times* for its usual dependability, he detailed his specific grievances. "I have just this moment seen the report on the front page of the Saturday issue, headed 'Millikan Retorts Hotly in Cosmic Ray Clash.' This article would be amusing in the completeness of its misunderstanding and consequent misrepresentation of my address were it not such misrepresentation, whether conscious or unconscious, has serious consequences."

Millikan said that at one point he happened to use the word "incautious" when referring to the cosmic ray energy measurements of some other experimenters, by which he meant "nothing caustic" and "not the remotest reference to Dr. Compton." Yet, Millikan complained, "the reporter, probably expecting that I would 'retort hotly to Compton,' apparently sensed a slur aimed at Dr. Compton, for that would make 'good news' and so built his whole report around this caustic adjective." This despite the fact that Compton had never made the kind of energy measurements to which Millikan had been referring, and therefore "was as remote from my thought as the man in the moon when I used the word."

But that wasn't the worst part, Millikan continued, accusing the *New York Times* of reporting the exact opposite of the actual findings he had presented at the meeting, specifically about the "complete agreement" between himself and Compton on the high-altitude cosmic ray latitude effects, and that "long before the newspapers had created a controversy over latitude effects," Millikan had been the person to undertake looking for such phenomena by airplane flights. "All this," according to Millikan, "the reporter apparently failed to hear."

The *New York Times*, in conclusion, had gotten it all wrong. There was no "clash," no great conflict. "Although our interpretations [of cosmic ray data] differ—as is altogether natural in any new field—the 'coolness' between Dr. Compton and myself was entirely in the reporter's mind, and there was not the remotest justification for such a statement as the *Times* report contained."[12]

That would have been news to the various other papers and magazines that specifically noted the tension between Millikan and Compton during the symposium, such as *TIME* magazine: "The audience sighed. Dr. Millikan sat with his arms clenched across his chest. Dr. Compton scowled darkly. Laureate Compton (to a questioner): 'Don't bother me!' Laureate Millikan: 'No controversy!'"[13]

"Millikan was protesting vehemently that I had done him a great injustice," Laurence remembered, noting that he hadn't written the headline about "retorting hotly," which had apparently hurt Millikan's feelings. "He lost his head completely and said he had never seen such misrepresentation."

The *Times* editors asked Laurence how they should respond to Millikan. Laurence didn't hesitate: publish Millikan's statement as is, he said. But he also claimed the right to respond in print.

He did so a week later. "I just took him to pieces,"[14] Laurence recalled. He stated outright that Millikan's interpretation of his article "is at variance with the facts," pointing out that the *New York Times* had published extensive verbatim quotations from both Millikan's and Compton's speeches. Much of what Millikan was complaining about hadn't even been addressed in the "Millikan Retorts Hotly" piece, "for the very reason that these two papers were dealt with separately in two special articles, one devoted entirely to Dr. Millikan's paper and the other exclusively to Dr. Compton's."

As to any misrepresentations of Millikan and Compton's "complete agreement," Laurence set the record straight using each man's own words, showing how their "opposite findings were clearly brought out, respectively, in the addresses by Dr. Millikan and Dr. Compton. . . . No two sets of facts could be more diametrically opposed, and it is on these very facts, the difference in intensity at sea level latitudes, upon which Dr. Millikan and Dr. Compton draw diametrically opposed hypotheses."

Laurence continued, "These differences of fact and hypothesis are known to the entire scientific world, and were, because of their importance, given prominence in my article, to make the entire subject clear to the lay public." And yet, "anyone reading Dr. Millikan's telegram might be led to believe that there is no fundamental difference whatsoever." Even if Millikan and Compton agreed on the facts about high-altitude variations, "they draw exactly opposite conclusions from these very facts."

As to the whole "incautious" issue and the accusation that Laurence had "built his whole report" on that "caustic adjective," Laurence remarked, "It so happens that the word 'incautious' was never used by Dr. Millikan and consequently never appeared in my article. Neither did the word 'caustic.' The exact phrase used by Dr. Millikan, and quoted by me, was 'less cautious,' a much milder term." He explained that considering the context, "a reporter may be excused for believing that the 'others, less cautious' included Dr. Compton, without being publicly pilloried," slyly referencing Millikan's own comment in his address about being "pilloried." In any case, even the term "less cautious" was mentioned only once, "so that in no conceivable manner was 'the whole story built around this caustic adjective.'"

Finally, Laurence wrote, his report of the "obvious coolness" between the two men was based "purely on observations of the attitudes of the two scientists toward each other." But if Dr. Millikan specifically denies it, then "I must offer my sincere apologies."[15]

There was no further word from Millikan—yet. Laurence, however, was pleased with himself and his handling of the matter. "In the scientific community I made a lot of friends because they were all there," he said later. "They heard it, you know. They saw Millikan's statement and saw what I had replied . . . they wrote me complimentary letters. . . . I got letters saying, 'That was a crackerjack of an apology,' because obviously there was coolness between the two men."[16]

Millikan had overreacted, not because of any inaccuracies in Laurence's reporting but largely because of the poorly worded headline that emphasized the "hotness" of his debate with Compton. In Millikan's mind, disputes among scientists were supposed to be handled in a gentlemanly fashion and most importantly, confined to strictly scientific venues instead of being splattered across the front pages of America's newspaper of record.

Laurence, however, knew better. Scientific method or not, unpassionate and impartial observers or not, scientists were just as human as anyone else, just as subject to envy, jealousy, and anger as they were to the more noble aspirations and ideals of their calling. He had, after all, spent many years reporting on the researchers, both distinguished and crackpot, who spent their time trying to "disprove" or "unseat" or "demolish" the theories of Einstein. He knew how contentious and petty scientists could be, and he wasn't about to disregard controversy or his responsibility to cover all sides of a subject just because a prominent scientist was unhappy that it might portray his profession in a less than noble light.

It would not be Laurence's final brush with Millikan. For now, at least, perhaps the sage of Pasadena had been mollified by a long *New York Times* Sunday supplement piece by science editor Waldemar Kaempffert that offered a somewhat more sedate and contemplative take on not only the cosmic ray wars, but also other subjects presented at the AAAS meeting than was apparent in Laurence's more immediate on-the-spot reporting. (The fact that Kaempffert also mentioned "a certain agreement between Compton and Millikan" might have also placated Millikan.)[17]

Now that the face-to-face, in-person clash had, by all estimations, ended in a draw, a sense of temporary calm set in. Millikan and Compton would each continue to examine the data they had collected, continue at a somewhat slower pace to gather more, and prepare their results for formal publication in the scientific press. The sages and scientists departed Atlantic City and gave the town back to the tourists, entertainers, and gangsters.

Not all of the big cosmic ray questions had been resolved, but along with the daily tides of the nearby sea, the coastal town of Atlantic City had witnessed the scientific tide decisively shifting against Robert Millikan and his romantic cosmologies, in ways that would continue to gather in strength and force. He knew it was happening and that it would eventually sweep away his treasured views of the universe, but he was also far too stubborn and determined to let it happen quite so easily.

13

HOORAY FOR THE COSMIC RAYS

As 1933 got under way, big things were happening on both coasts. The irrepressible Auguste Piccard arrived in New York City to visit his twin brother Jean Felix, embark on a speaking tour about his stratospheric adventures, and drum up support for his next adventure in the frozen North. On the West Coast, the far more restrained and ethereal Albert Einstein arrived back in Los Angeles, making his fourth trip to the United States and his third consecutive New Year's visit to Caltech.

Robert Millikan was back from Atlantic City in time to meet him, posing for pictures with Einstein and his wife as they disembarked from their ship, the steamer *Oakland*. As usual, Einstein was greeted by a passel of reporters firing questions both provocative and inane. Again as usual, Einstein was masterful in dodging the latter and responding only to a select few of the former.

Inevitably, one reporter tried to draw Einstein into the Millikan-Compton debate that had been dominating the newspapers. Did Einstein think that cosmic rays were photons, as Millikan held, or were they particles, as Compton insisted? And did a latitude effect exist? Einstein merely responded with one of his trademark enigmatic smiles and pointed his pipe at Millikan. "There is a man better able to talk about cosmic rays right over there," he said.[1]

13.1 Millikan (right) greets Einstein and wife upon their arrival in California, 1933. (Wikimedia Commons)

As he had the previous years, Einstein would spend about two months in California, meeting and discussing with colleagues matters of cosmology and physics, giving speeches and press conferences, and in general being Albert Einstein.

Arthur Compton was back in Chicago, biding his time, confident in his position, and still collecting data from his remaining survey expeditions. As with Millikan and his falling balloons and wonky electroscopes, not everything would always go well. In Santiago, Chile, one of Compton's researchers, Paul Ledig, fell victim to petty crime, losing some cosmic ray photographs in the process. "Mr. Ledig had taken off his coat and left it lying on the grass," according to an Associated Press report. "The thief took his wallet containing the photographs and a small amount of money."[2]

Meanwhile, in Stuttgart, Germany, Erich Regener was launching more balloons, reaching nearly seventeen miles in altitude and capturing more cosmic ray images.[3] The Millikan-Compton controversy may have been in a holding pattern for the moment, but the research continued, even by those with no personal stake in anything but the scientific truths to be determined. As the experimenters labored on, the theorists continued to theorize.

One of them was Piccard, although he was also decidedly a member of the experimental camp, as his daring stratospheric flights had demonstrated. After meeting with President Hoover and lunching at the Belgian embassy in Washington, D.C., he spoke at the National Geographic Society, where he declared that cosmic rays could provide the world's main energy source in the future. "And when this source of power is made available, producing limitless energy almost free, what coal remains in the earth's veins will suddenly become valueless."[4]

It was nothing really new, as talk of "limitless energy" from some marvelous type of atomic power had already been going on for some time, including the idea that cosmic rays would somehow be involved. But Piccard's enthusiasm made it all fresh again, despite the fact that the two recognized preeminent cosmic ray authorities, Millikan and Compton, had generally dismissed such wild speculations in the past about their pet subject.

Anyone discussing cosmic rays naturally invited comparison and comment on the Millikan and Compton contretemps. But while some reports interpreted Piccard's remarks that "somewhere along the Milky Way in the celestial laboratories these rays are flying off from the very process of creation" as support for Millikan's birth cries ideas, Piccard refused to take sides.

As always, however, the *Los Angeles Times* was quick to jump on any possible snippet of Millikan boosterism. "August [sic] Piccard rushes over to America all out of breath to tell us that when coal and iron deposits fail, we may be able to obtain enough energy from three drops of water and cosmic rays . . . and California has the big cosmic ray chief in Pasadena, Dr. Millikan!"[5]

Even in the midst of an economic depression, not everyone was necessarily enthusiastic about the notion of unlimited energy from cosmic rays. "Great! Splendid! Hooray for the cosmic rays!" exclaimed a humorist in the *Chicago Tribune*. "But wait a minute. That will throw all the coal miners out of jobs, won't it? Not so good. There's enough been put

13.2 (Left to right) Robert Millikan, Georges Lemaitre, and Albert Einstein at Caltech, 1933. (Wikimedia Commons)

out of jobs by oil burners. That's the trouble with these scientists. They can create all sorts of new things except jobs."[6] Another commentator penned a verse apparently attributing mind control powers to cosmic rays: "Though a citizen and voter/I deserve no blame or praise;/I am just a human motor/And I'm run by cosmic rays."[7]

Less flamboyant but more dignified than Piccard was Abbe Georges Lemaitre, back in Pasadena to speak on his own cosmological ideas and to ponder them further with Einstein and the Caltech-Mt. Wilson Observatory brain trust. Again, he mentioned that he hoped to broker an intellectual peace between Millikan and Compton. In a talk at the Mt. Wilson library at which Einstein sat calmly as just another member of the audience, Lemaitre again expounded about his "primordial atom" and how its explosion, or sudden expansion, about ten billion years ago formed the current universe.

His ideas, in fact, could be interpreted as pro-Millikan, pro-Compton, or even a synthesis of the theories of both. He remarked that "cosmic rays are some kind of a glance preserved for us in ever-increasing space of the primordial fireworks that marked the beginning of the universe." Did that mean they represented universal birth cries? Not so fast: "I believe that cosmic rays provide us with the sole proof of the former existence of the super-radioactive materials whose gradual disintegration gave birth to them. . . . The cosmic rays we now study must have been many times more penetrating at the distant time at which they were born out of the decomposition of the super-radioactive atoms."[8]

Beginnings and disintegration and decomposition: cosmic rays were, according to Lemaitre, a little bit all of the above. In effect, they were the energy left over from the creation of universe itself, now very slowly running down as it continued to inexorably expand into infinity. "But it is the study of cosmic rays in which Dr. Millikan and Dr. Arthur Compton have been engaged that will give us the final answer as to how the universe was created," Lemaitre explained. "I really think that the study of cosmic rays will give the final answer to cosmological questions, that they are the key to the problem."[9]

Einstein, for one, was impressed. "It is the most pleasant, beautiful, and satisfying interpretation of cosmic radiation I have listened to," he said. "This picture has fewer objections and conjures less contradiction than any other theory of the cosmic ray source." Unfortunately, neither Millikan nor Compton were present to hear Lemaitre's talk, so any opportunity for his ideas to form grounds for reconciliation would have to wait until another time.[10]

Perhaps Millikan passed up attendance because of Lemaitre's just-published paper in *Physical Review* with MIT collaborator M. S. Vallarta, which used Vannevar Bush's "mechanical brain" at MIT to model the effects of the Earth's magnetic field on charged particles. The paper reported that the data collected by Compton's world survey that found "a close correlation with magnetic latitude" not only agreed with that gathered by other investigators such as Jacob Clay, but also was precisely the effect predicted by their model of the Earth's magnetic influence. "This discovery rules out the hypothesis that the cosmic radiation consists of photons alone and suggests that it is made up at least partly of electrons, protons or other charged particles," Lemaitre and Vallarta asserted. "Compton's result definitely shows that the cosmic rays contain charged particles." The latitude effect was proven not just with electroscope data collected from far-flung exotic locales by intrepid researchers, but also by cold, hard mathematics.[11]

Still, while quiet, learned talks in academic libraries about elegant and beautiful theories were all very nice, it was clear that the answers to the outstanding questions regarding cosmic rays, and with them perhaps a final settlement of the Millikan-Compton controversy, would come only through hard-won, crystal-clear experimental evidence. The efforts to collect that evidence had hardly slowed down. And both Compton and Millikan remained right in the thick of it all. Millikan would now begin to try to find a way to cede some ground to Compton while also maintaining that in other ways, he was still right and in fact had been right all along. If there was to be a reconciliation, Millikan was determined that it would be on his own terms.

In early February, now armed with the fresh data from Victor Neher's readings taken on his high-altitude airplane flight in Panama on his return trip home from Atlantic City, Millikan began to admit the existence of a cosmic ray latitude effect—sort of. Preparing to publish a revised version of his Atlantic City AAAS presentation in *Physical Review*, he announced that Neher's new data confirmed Compton and Clay's reports of at least a high-altitude latitude effect, with a difference of as high as 8 percent lower in equatorial regions than at the poles.[12]

Why hadn't he seen this before in his own expeditions dating back into the 1920s? The simple explanation: he and his collaborator Cameron actually *had* noted the effect but had dismissed it as an anomaly because "these readings were taken before we had developed sensitive high-pressure electroscopes and we were unable at that time to lay much claim to precision," along with several other difficulties that caused them to discount the readings they had made on and under water at high-altitude mountain lakes. The next step should be to gather more data with unmanned balloons at equatorial regions to the highest altitudes possible, Millikan said.[13]

It seemed a way for Millikan to have his scientific cake and eat it too. As De Maria and Russo summarized the situation, "Millikan did observe the latitude effect before Clay [or Compton], but he chose not to recognize its existence because it did not fit his theoretical preconceptions," meaning his cherished ideas of cosmic ray energy bands and hence the birth cries theory they ultimately seemed to ordain. They offered another caustic evaluation: "Compton's students put the point more neatly. They synthesized Millikan's pleadings as follows: 'In the first place I don't believe that there is a latitude effect but if there is I discovered it.'"[14]

To some press observers, this seemed to herald peace in our time, at least as far as the two Nobel giants were concerned. "To honest Dr. Robert Andrews Millikan, a fact is a fact," declared *TIME* magazine. "Although it fortified the cosmic ray theory of Dr. Arthur Holly Compton, who had become Dr. Millikan's opponent in the field of cosmic ray theories, Dr. Millikan at once published his new finding. . . . Caltech's Dr. Paul Sophus Epstein, mathematical physicist, patted the proud heads of both scholars last week. . . . Nobel Laureates Millikan & Compton are both correct in their theories, testified Dr. Epstein." That was possible, said Epstein, because cosmic rays basically came in two varieties in about a 70/30 proportion: a hard, highly penetrating component of photons, which penetrate all the way to the ground even at the equator, and a softer, less penetrating component of particles that are diverted by Earth's magnetic poles. Such a conception was also apparently consistent with Lemaitre's theories.[15]

The *New York Times* also seemed to agree, observing that "the paths of Dr. Millikan and Dr. Compton were leading in the same direction, and . . . eventually the varied opinions of the two ray hunters would merge."[16]

Despite such hopes, however, the battle was far from over. To the contrary, it was about to take center stage again in an even more conspicuous, spectacular, and public arena than the relatively staid setting of a scientific conference: the shamelessly unabashed spectacle called a World's Fair.

Millikan and Compton may have been the cosmic ray superstars as far as the press and public were concerned, but other scientists continued to do important work in the field, not only in Europe but in the United States as well. One of them was W. F. G. Swann, director of the Bartol Research Foundation, who among his many balloon experiments and other research even managed to snare some press coverage once in a while, especially when he did things such as locking himself in a steel vault far beneath a Manhattan office building to demonstrate the power of cosmic rays. Swann used a Geiger counter with an amplifier and microphone setup to broadcast the sound of their detection over seventy radio stations, giving cosmic rays a huge audience.[17] In an era of daredevils, flagpole sitters, and other record setters, Swann's stunt was fairly modest but enough to attract some minor attention.

A much bigger spectacle was coming together in the ongoing cosmic ray circus, however. Not surprisingly, Auguste Piccard was again involved. Continuing his lecture tour of the United States, Piccard stopped off in the Chicago area to speak at Northwestern University in mid-February and then the University of Chicago at the beginning of March, where he was introduced by Compton. The men also met as scientific colleagues to discuss their mutual interest in cosmic rays, Piccard telling the press that his own studies complemented those of Compton and that the Millikan-Compton controversy "will soon be settled," although he didn't explain exactly how that would happen.[18]

The apparent answer would not be long in coming. The Century of Progress International Exposition, more commonly known as the Chicago World's Fair, was set to open at the end of May. The fair organizers had already been thinking about a balloon flight, and Piccard's visit to the United States provided further inspiration.

"Auguste's fame, as well as his lectures and attendant publicity certainly helped to convince the planners of Chicago's 'Century of Progress' exposition to include a stratospheric balloon flight as a dramatic demonstration of how men of science and men of industry could work together to extend the boundaries of useful knowledge," explained historian David DeVorkin.[19] A balloon spectacular would also encourage industrial concerns such as Dow Chemical and Goodyear to donate support and funding, not only to the Chicago World's Fair but perhaps even to additional Piccard flights.

While Auguste and Jean Piccard were in town, they met with Compton and World's Fair officials to begin thrashing out plans. Compton was the natural choice to handle the scientific aspects of the flight and was happy to take advantage of the opportunity to get further high-altitude data without the effort and expense of having to again traipse his equipment up mountaintops. Here were the Piccards offering to carry his instrument up into the sky from Compton's own backyard, South Chicago. There was nothing to lose.

Compton apparently surmised that his archrival Millikan, then embarked on another campaign of high-altitude sounding balloon and airplane measurements, would feel similarly. "As you may have heard, the Piccard brothers are planning a high altitude balloon flight from Chicago some time during the latter part of June or the early part of July," he wrote to Millikan in early May. "It would seem too bad to let an expensive flight of this kind occur without making use of it for some high altitude measurements." Compton explained that the World's Fair committee in charge of the flight had asked him to get together cosmic ray equipment to take along, noting, "It has occurred to me that it would make an interesting check on your sounding balloon measurements if you would care to send up with this balloon one of your new recording electroscopes, such as is described in the last number of the *Physical Review*."

Perhaps to give Millikan a little more incentive to participate, Compton remarked that if Millikan wasn't able or willing to supply one of his own electroscopes, then Compton's group "shall probably desire to build an approximately equivalent one to send with the balloon."[20] Given the enthusiasm with which Millikan had been boasting about his electroscopes in the press and scientific literature, Compton was undoubtedly confident that the thought of sending up a pale imitation would be anathema to him.

Millikan responded a week later, saying, "We shall be glad to send up one of our latest type self-recording instruments in Piccard's balloon if you desire to have it go along," as if he were doing Compton a great favor in deigning to agree. He emphasized that no special technical arrangements would be needed for his lightweight apparatus and that because it could operate all on its own, "Dr. Piccard [then assumed to be piloting the flight] will not have to do anything about it whatever."[21]

Again the two cosmic ray giants would meet, or at least their instruments would, coming together in common cause to celebrate science and progress at the World's Fair. "Compton saw an opportunity to settle his debate with Millikan through another type of collaboration," wrote DeVorkin. Not to mention the Piccard balloon flight raised the possibility of demonstrating whose electroscope design was better, Compton's or Millikan's, an issue that had become a major concern for their mutual funder, the Carnegie Corporation.[22]

The Millikan-Compton feud had evolved into its next stage, in which rather than two heavyweights clashing in the ring for the world's championship title, it was a more-or-less noble striving for mutual understanding between two dignified diplomats. And that understanding would be achieved not through conflict, but through the progressive and exalted means of science and peaceful effort, the most noble aspirations of humanity.

Or perhaps not. There were certainly more than enough parties interested more in conflict and confrontation than in peace and pacifism, such as the editors of the *Los Angeles Times*, who continued to play the role of Millikan's most enthusiastic and vocal media cheerleaders. Commenting on a recent Caltech lecture, they announced, "Dr. R. A. Millikan heard his theory as to the nature and origin of cosmic rays substantiated by one of his boys, R. M. Langer, who showed that such rays cannot consist of charged particles born in the upper air—as claimed by Prof. A. H. Compton. . . . Recognized as [an] outstanding physicist, Prof. Compton's hurried adventure into cosmic ray regions still has his colleagues guessing. His announcements were boldly broadcast yet have been very weakly supported, and his admirers suspect he may still have a few facts up his sleeve." Perhaps Compton would let loose his "bolt" at the next AAAS meeting, speculated the *Los Angeles Times*, though "whether a thunderbolt or popgun remains to be determined."[23]

In his public lectures, Millikan himself didn't do much to extend any fig leaves to Compton, calling the idea of cosmic rays originating in the upper atmosphere as "crude and unintelligent" and "foolish talk."[24] The man who "makes the cosmic ray turn handsprings and jump through hoops" also took a shot at some experts outside of his own field: "I have been trying for years to find a field of economics upon which we all agree, and I find that economics is all fringe," proclaimed Millikan. "The economists agree on nothing."[25]

As preparations for the World's Fair picked up in Chicago, Compton had the opportunity to exchange deep thoughts with Einstein, who was stopping off in Chicago after leaving California. Einstein again voiced his support of Lemaitre's ideas and, it seemed, also for Compton's take on cosmic rays.[26] On his way back home, Einstein would be confronted with more immediate and far less cosmic concerns: while he had been away in Pasadena, Hitler had taken absolute power in Germany, and the Nazis had been raiding his home and seizing property. Einstein would renounce his German citizenship and eventually settle in the United States two years later—but in Princeton, not Pasadena.

As Millikan and Compton continued with their own individual research programs, the World's Fair was coming up. Because of his close geographical proximity to the proceedings, Compton likely felt more of a sense of urgency, watching the fairgrounds on the Near South Side lakefront being cleared and landscaped and prepared for the festivities and the various buildings and exhibits under construction.

One of them would be a "great temple of science," the *Chicago Tribune* proclaimed, a place where "one may bow at the gates of the invisible forces which rule the universe." Of course, some of those forces definitely involved cosmic rays, "which may prove the source of a new power that can be harnessed into machinery, power taken from the air to make light for humans and to run automobiles." To illustrate such fanciful ideas, "cosmic rays will be utilized to produce the electric current which will informally set the assembly machinery of the General Motors exhibit in motion," in a demonstration conducted by Compton himself and presided over by General Motors president Alfred P. Sloan Jr.[27] Another cosmic connection would be provided by starlight from the red giant star Arcturus, which would be captured at nearby Yerkes Observatory and used to activate the World's Fair lights on opening night.

Other exhibits included "a mechanical man, made to look like a human being, 10 feet high" that would "give twenty minute lectures on food chemistry and nutrition." Fairgoers would be able to watch the robot demonstrate the digestive process as he traced food through his own stomach and alimentary canal. "This will be open to view and there will be a movie in his stomach, showing digestion, and how the fluids come from the walls of the stomach to do their work."[28]

It is unlikely that the digesting robot displayed the workings of the alimentary canal all the way through to the end, but that particular function became a concern for fair officials as opening day neared. "About the only thing that perturbed the Fair officials yesterday was complaints from some of the concessionaires on the comfort situation," the *Chicago Tribune* reported. The concessionaires "charged that the comfort facilities were not only inadequate, but too many of them were to be operated on a 'drop a nickel in the slot' basis."[29]

Plans for the Piccard balloon spectacle were running into their own snags. Although the experienced U.S. Navy balloonist Lieutenant Commander Thomas G. W. "Tex" Settle had been named as pilot by Auguste Picard, it was still undecided whether brother Jean or Auguste, whose enthusiasm for aerial adventuring had diminished, would accompany Settle. As the gondola was being prepared, Compton and Millikan were thrashing out the details of the cosmic ray equipment that it would carry aloft. It was proving to be somewhat more complicated than previously anticipated.

For one thing, Millikan was distracted back in Pasadena playing host to another world-famous savant, Danish genius Niels Bohr, along with all his other work and responsibilities. Naturally, the *Los Angeles Times* enthused that Bohr was in Pasadena "to go into huddle with Drs. Millikan, Von Kármán and Paul Epstein of Caltech over the cosmic ray." That was perhaps one of Bohr's intentions, but hardly the only one. Still, the *Los Angeles Times* was confident that "the entire matter is in good hands and we are content to struggle over a raise in wages instead of cosmic rays."[30]

Compton also experienced a potentially greater distraction in mid-May when he totaled his car en route to a brief college lecture tour in Iowa. The car skidded on wet pavement near Sterling, Illinois, and hit another car, injuring both Compton's wife and son Arthur Alan in the process. Fortunately, nobody was seriously injured, and Compton was able to continue on to do his lectures while his slightly battered family returned home to Chicago.[31]

Such complications aside, Compton and Millikan had to get everything ready if their instrument proxies were going to take flight at the World's Fair. The Piccard flight was tentatively scheduled to take place on or around July 1, depending as usual on weather, always a fickle phenomenon in Chicago.[32]

As the scientist on the scene and in charge of the instrumental setup, Compton had to find ways to make both his and Millikan's instruments work together and more or less automatically, with little or no attention from the balloon crew. Impressed with the automatic recording mechanism of the Millikan-Neher electroscope, he cabled Millikan to ask whether he could supply an identical recorder that could be used on Compton's electroscope. Whether out of some sort of personal resentment or jealousy, or simple technical infeasibility, Millikan declined: "Cannot supply new recording mechanism capable of adaptation to other types of electroscopes in the ten days specified," he wired Compton. "Regret shortness of time makes furnishing recording mechanisms impossible."[33] Compton, it seemed, would have to supply his own recording equipment.

No matter, Compton gracefully responded in a letter. "I of course understand the difficulty in supplying us with a recording camera on such short notice," he wrote. Again, he mentioned that he was "much pleased indeed" that Millikan would be able to participate, adding that he hoped "the work on the Piccard balloon will have progressed far enough when you are here in June so that you can yourself supervise the installation of this equipment in the gondola."

Perhaps a bit waggishly, Compton also enclosed with his letter a *Chicago Daily News* clipping titled "Piccard Flight May End Compton-Millikan Debate on Cosmic Ray Properties." Compton remarked, "I may add that much as we naturally would like to avoid publicity, in connection with these experiments, I suppose in the present case it is only by virtue of some publicity that the cost of the flight can be met. If the News correspondents continue to write in the vein of this article, it will not be so bad."[34]

Compton was certainly right to point out that as far as the press was concerned, Chicago had been definitively identified as the next arena of battle in the great Millikan-Compton contest. The Associated Press explicitly said as much in its report, declaring that Millikan and Compton "have been granted a change of venue in their famous dispute which involves the cosmic ray. . . . When a balloon bears one of the Piccard brothers on another flight into the stratosphere in July it may serve to quiet for all time this ever-flaming friendly dispute between these two Nobel Prize winners."[35]

The joint nature of the flight, however, with its implied cooperation between Millikan and Compton—friendly or not—seemed to somewhat muddy its public portrayal. Whereas the Atlantic City debate had been widely depicted in terms befitting a heavyweight prize fight, the World's Fair experiment was more difficult to characterize. The Associated Press's use of the term "change of venue," for example, could easily be interpreted as a reference to the legal practice of transferring a court action to a new location for trial. The *Chicago Daily News* article cited by Compton tried to have it both ways, describing how the "disputants" Millikan and Compton would now take their differences to the stratosphere for a "final hearing." But writer Dempster MacMurphy also deployed a sports analogy by pointing out that "the cosmic ray itself [will act] as referee."[36]

Unfortunately, the Millikan-Compton feud, fight, debate, or whatever it might be called, was not about to be settled quite so easily or neatly.

14

FALLING AND RISING

As previously arranged and publicized, the lights of the Century of Progress International Exposition were activated on Saturday, May 27, 1933, as light from Arcturus was focused on photoelectric cells at Yerkes Observatory. Two days before, the General Motors exhibit had been dedicated and set into motion by Compton's cosmic ray detector. As Charles Kettering, the corporation's vice president in charge of research, spoke to a group of "leaders in science, education, industry and politics" assembled for the occasion, "cosmic rays, about which we know so little, were flashing lights on both sides of the dining hall, discharging positive and negative electrons as they crashed into gas atoms at a rate of 186,000 miles a second," reported the *Chicago Tribune*.[1]

The fair might be open and running, but neither Piccard's balloon nor the cosmic ray instruments it was to carry were ready. Compton and Millikan were still thrashing out the details, while Millikan was continuing to send Victor Neher on high-altitude flights from March Field with his "famous instruments designed to trap cosmic rays."[2] In case anyone doubted the importance of such work, the *Los Angeles Times* was happy to set them straight: "This cosmic ray thing may sound like a lot of bunk to some people. But you never can tell. We may all be running our refrigerators and autos with them yet; perhaps they will warm our houses and sweep our floors. Nothing can seem too fanciful after radio's marvels."[3]

Millikan and Compton were also both preparing for the summer American Association for the Advancement of Science meeting in Chicago, being held in conjunction with the World's Fair. As Compton had mentioned in his earlier letter, he was hoping that while in town for the AAAS meeting, Millikan might be able to personally supervise the installation of his equipment, but that proved impossible. Problems with Piccard's gondola had pushed the flight back from early July to at least the middle of the month. That was not a huge difficulty, since the World's Fair would be humming along well into 1934, and there was still plenty of summer left before the Chicago fall and winter would set in.

Perhaps to preserve their newly acclaimed cordial and cooperative relationship as manifested in the upcoming Piccard flight, at this AAAS meeting Millikan and Compton consciously avoided the charged and contentious atmosphere of six months earlier in Atlantic City. There was no official "debate" scheduled, merely each man calmly presenting his latest work. For Millikan, however, that meant yet another opportunity to present his ideas of cosmic creation. Announcing that the mass of the newly discovered positron had been determined by Carl Anderson, along with the results of his latest high-altitude cosmic ray airplane flights, he held that this latest evidence served to further support his theories.

"It seems to be becoming popular now for the astronomers to use this synthesis hypothesis instead of the annihilation hypothesis to explain the evolution of heat energy by the stars," he said. "Indeed, the annihilation hypothesis"—as championed by Jeans and others—"seems at present to be in a state of eclipse. The question may then be raised whether synthesis can explain both the cosmic rays and stellar energy. There is no reason why it may not be called upon for both purposes, but with a different mechanism." Compton refrained from any direct discussion of cosmic rays in his own presentation, but instead discussed how the discovery of new particles could provide fresh insight into the structure of the atom.[4]

Interestingly enough, even at this late date, Millikan was still being credited in some circles as the "discoverer of the cosmic ray," despite the fact that any such claim had long been disproven. His eminence and reputation continued to cast a very long shadow.[5]

Journalist William Laurence, who had become an inadvertent footnote in the Millikan-Compton feud six months earlier, would discover yet another side of the Millikan personality in Chicago. He recalled, "I was sitting in the press room; there were several other newspapermen there; and who should walk in but Millikan. There was a great silence. That was the first time he and I had seen one another since the fracas [in Atlantic City], and I thought, 'My God, he is going to pounce on me.'"

All waited for an inevitable explosion, but Millikan kept his cool. "Come on, Laurence, I want to see you," he said. "Come on outside." A nervous Laurence followed him out of the press room.

After a silence, Millikan said, "Laurence, I'm sorry what happened. I realize now that it wasn't your fault at all."

Before a stunned Laurence could say anything, Millikan's eyes flashed in anger and he pointed a finger at Laurence. "It was all Compton's fault!" he declared.

"So that was the end of a famous episode," Laurence remembered later. "That was the only time that there was any misunderstanding between me and any scientist about any report in all the 34 years I reported science on the *Times*. And even that one . . . the scientist apologized to me."[6]

Millikan had not, of course, apologized to Compton, for Atlantic City or anything else, nor did he have any intentions to do so anytime soon. Neither did Compton have any such plans. Quite to the contrary: while they awaited the latest Piccard balloon excursion, both men were busily planning their next steps in their campaigns to decisively win the argument at last.

Perhaps in an attempt to emulate or even supersede Compton's worldwide survey, Millikan was arranging to expand his own global database of cosmic ray data. He began to contact cruise ship lines, beginning with the Grace Line on which Victor Neher had sailed on his earlier South American trip, requesting to place his recording electroscopes on their ships. "We are making a world-wide survey of these cosmic ray intensities and have developed recording apparatus which gives us automatically records which we need," he explained. Millikan was especially keen to get an electroscope on a vessel making a round-the-world voyage. He emphasized that his equipment would not need any special attention aside from "winding up, say, once every two weeks, much as a clock is wound up."

The ship lines replied that they were happy to cooperate, providing schedules of the vessels that would soon depart from Los Angeles. The automatic nature of Millikan's gear was undoubtedly a good selling point. "The Master of the ship would attend to having the clock wound up at regular intervals in accordance with instructions. This could be handled by the ship's officer who is entrusted with the duty of winding the ship's chronometers."[7]

Meanwhile, another balloon flight was being quietly planned. Millikan received a letter from Captain Albert W. Stevens, informing him that the Army Air Corps was preparing to make another try at the high-altitude record. The commander of the Air Corps had ceased such flights after the 1927 death of an officer from hypoxia in an open basket, but matters had changed, said Stevens: "The new Chief of Air Corps . . . has a different attitude and gave official consent last February to a high altitude flight in an inclosed car . . . to an altitude as high as it seems physically possible to go."[8] Although Stevens didn't express it specifically in his letter, it's likely that the army was looking to stake its own fresh claim to the stratosphere, rather than cede it to the navy's balloons and airships or to foreign adventurers such as the Piccards.

Stevens was no stranger to such exploits, an experienced balloonist and aerial photographer. He acknowledged the upcoming Settle-Piccard World's Fair flight and that "Dr. Compton has contributed largely to the project, that the flight probably will give him considerable publicity, and that his apparatus is to be carried in the gondola," and that Millikan would also have equipment on board. The Air Corps flight, however, would use a balloon five times larger than Piccard's, which would already be the largest balloon yet flown, and thus be capable of besting whatever record might be set in Chicago. Although with Piccard's 600,000-cubic-foot balloon "an elevation of 55,000 feet may be expected," the Air Corps' planned three million-cubic-foot balloon "should go 20,000 feet higher than the one to be used in Chicago," according to Stevens. "On the flight we have in mind we would like to carry at least two instruments, one or both of which may be yours," he wrote.[9]

Stevens also inquired whether Millikan could get funds to help pay for the project. "It should be understood that the Air Corps, because of reduced appropriations, cannot pay for the balloon or the gondola, but can furnish practically everything else." Stevens explained that he was approaching potential contributors, including the National Geographic Society, but "we would like to keep the number of contributors down to a minimum and to secure several rather large appropriations instead of many small ones. We hope to keep this a purely scientific project, and do not wish to get mixed up in any advertising schemes or undesirable publicity." For that reason, "we have purposely refrained from any publicity until the Chicago flight is over, and even then we do not propose to say much about our own flight until equipment is ready."[10]

The flight would depart from a remote location "where the public can be kept at a safe distance" without any possibility of "interference by the public, newspaper men, motion picture operators or others." Such "public pressure," Stevens thought, might cause problems with the Chicago flight by forcing it to launch under less than perfect conditions.[11]

Millikan responded with somewhat guarded enthusiasm. "We shall be most happy here to assist [the flight] in every way we can, and in particular to provide and take charge of tests to be made with recording cosmic ray electroscopes." He made it a point to note that "the techniques which we have developed here and have had in continuous use for more than a year, I think it would be generally agreed, are greatly superior to anything that has been done elsewhere."[12]

His support, however, was slightly qualified, because "the most vital thing for our understanding of these rays is that these high flights be made not here in the temperate zones, but in the equatorial zone." In other words, Millikan would provide whatever help possible for the Air Corps flight, but he would really prefer it to be done at a different latitude.[13]

In between making preparations for the World's Fair flight, Arthur Compton was also exploring new possibilities. He had enlisted famous explorer Admiral Richard E. Byrd in his cosmic ray campaign, announcing that Byrd had agreed to take along a Compton cosmic ray "trap," a duplicate of the one about to go upward in Chicago, on his next expedition to the Antarctic later in the year. Accompanying the instrument would be physicist Thomas Poulter of Iowa Wesleyan University, already scheduled to go on the expedition to study meteors in the polar regions. "Admiral Byrd is going to study cosmic rays on his next trip to the Antarctic, but he isn't going to bring any back with him," remarked the *Boston Globe*.[14]

Far from the South Pole, at Soldier Field in south Chicago, the balloon was about to go up.

In an era when high-altitude ballooning had the same cachet as space travel would have several decades later, the men (not women just yet) who piloted those vehicles were the nearest thing to Alan Shepard, John Glenn, and the others who would make America's first manned spaceflights. And in 1933, U.S. Navy Lieutenant Commander Thomas G. W. "Tex" Settle was the closest thing to an astronaut, with a background equivalent to those who would fly on Mercury, Gemini, and Apollo thirty years later. A Naval Academy graduate, airship pilot, test pilot, and award-winning balloon racer, he was the perfect choice to take the *Century of Progress* balloon, as the craft was officially dubbed, into the stratosphere.

But even as Settle prepared meticulously for his upcoming adventure, supervising and testing the equipment at balloon-maker Goodyear's plant in Akron, Ohio, trying out submarine rescue breathing gear to use in case

14.1 Before the launch of the *Century of Progress* balloon, pilot Lieutenant Commander Tex Settle poses for newsreel photographers. (Wikimedia)

of emergency oxygen loss, and even spending eight hours locked inside the seven-foot-diameter metal ball of Piccard's gondola to check its habitability and ensure its airtightness, two questions remained.[15] Who would accompany Settle into the sky as copilot/observer, and when would the ascent actually take place? The answers kept changing.

Originally, the fair organizers had invited Auguste Piccard to be the star of the enterprise, but having lost his taste for such adventures, he demurred in favor of brother Jean, also an experienced balloonist. When Jean's lack of a U.S. flight license was pointed out, Auguste decided that Settle would pilot while Jean went along as observer, and such was announced to the press.

But matters weren't quite as settled as they were portrayed. There was still the possibility that Compton, officially in charge of the scientific aspects that were supposedly the mission's entire raison d'être, would decide to send along one of his own team to monitor the cosmic ray instruments. There was also a desire among some fair officials to keep the whole enterprise strictly an all-American effort. By this point Auguste had returned home to Europe and was serving merely as an advisor to the Chicago flight, while his temperamental brother Jean had been creating various complications.

Finally, as weather conditions at last seemed to be coming together for a launch in mid-July, the fair issued an official statement: "Dr. Jean Piccard today offered to remain behind as ground pilot of the Piccard-Compton stratosphere ascension and permit Lieutenant Commander Settle to soar alone." The purported reason was to save weight and thus make it easier for Settle to achieve the highest altitudes possible and a new world's record. The decision would "write a fine chapter of gallantry into the history of the flight," said World's Fair officials, because it "virtually guarantees [Settle] an altitude topping by thousands of feet the record that Auguste Piccard now holds." It was apparently not a sudden decision, since Settle had already been receiving training since at least a week earlier in operation of the cosmic ray equipment.[16] Jean had actually been squeezed out of making the flight by the officials, who considered him untrustworthy and erratic.

Assuming that Auguste Piccard's record was indeed broken, the Piccards could still take solace that it had been made possible by their expert counsel, not to mention Auguste's design of the Dow Chemical-built gondola. That was made obvious by the "PICCARD-COMPTON STRATOSPHERE ASCENSION" legend emblazoned on the craft in large letters—although by the time the balloon actually went up, that had been changed to "A Century of Progress," after the fair.

The question of the crew complement aboard the gondola, if somewhat slightly confusing to outside observers, was still wholly under human control. The issue of weather, however, was not. Weather forecasting was still a fairly inexact endeavor in the early 1930s in the days before computers and satellites, more of an art than science, but just as crucial to any sort of flying then as in the twenty-first century. As a veteran aviator, Settle was an expert meteorologist and kept close watch on conditions, monitoring a high-pressure area in Canada that would hopefully arrive in Chicago soon and create an ideal opportunity for liftoff.

Settle, Piccard, Compton, and everyone else anxious for the flight would have to wait a little longer on the vagaries of Midwestern summer weather. The expected high-pressure area fell apart, storms developed to the south and east, and Settle's launch window disappeared. It wasn't merely a matter of clear weather at the launch site: he also had to ensure that conditions to the east, to which the balloon would drift, were favorable, not to mention the winds at the varying altitudes through which he would pass.

While Compton's and Millikan's painstakingly designed and assembled instruments awaited their journey, and their builders waited to see whether the balloon flight would indeed settle their feud at last, the two scientists explained the importance of the whole enterprise to the public. In several syndicated newspaper articles, Compton and Millikan separately laid out the case for cosmic ray research and the reasons that men such as Settle were willing to risk their lives in its cause. "Why do men risk both life and limb in these perilous ascents? What urge tempts men to stray from the surface of the earth—to fly 10 miles or more into the heavens . . . ?" Compton wrote. "The answer is, briefly, that we see in a detailed knowledge of cosmic rays a possible answer to the age-old riddles of how this world of ours came to be."[17]

While Compton's articles were largely general and impersonal in their brief description of the state of the cosmic ray art, Millikan's contribution was far more personal, self-referential, and self-justifying. Where Compton used the "I" pronoun exactly once, while also including several mentions of Millikan's contributions to the field, Millikan's article is sprinkled throughout with "I" and descriptions solely of his own work, without any acknowledgment of others. From the beginning, it is a personal statement rather than an objective review. "I have been asked to explain why I am interested in making accurate physical measurements in the stratosphere, and I am very glad to do so," he wrote. As he had done earlier in the *Physical Review*, he depicted the current controversy as a minor, inconsequential difference of opinion. "The precise nature of these rays is not important," he wrote, despite the fact that it was the entire crux of the controversy. It didn't matter, though, he claimed, because "we have already obtained much excellent evidence upon this point" and agree on many aspects. Millikan apparently took the invitation to pen an article as another opportunity to push and restate his own positions to a fresh audience. "If all goes well, these observations should furnish interesting checks on the accuracy of our airplane and sounding balloon flights," Millikan noted, as if the most important purpose of the upcoming balloon flight was to serve as a check on his own work. That, of course, was the carrot Compton had advanced when he invited Millikan to participate in the project, and clearly it had been more than effective.[18]

Just as the tantalizing prospect of a resolution of the Millikan-Compton argument seemed only a balloon flight away, another possibility, or perhaps complicating factor, arose from one of their colleagues. Dr. Thomas W. Johnson, assistant director of the Bartol Research Foundation and one of the less-noticed participants in the Atlantic City debate, had been busy down in Panama making his own cosmic ray observations. In a talk to the Panama Canal Natural History Society, he announced that his work apparently settled one outstanding cosmic ray question: at least 30 percent of the rays were not electrons, as many (including Compton) had opined, but were positively charged, either protons or alpha particles. The rest were probably photons, as Millikan held. The conclusions were proven by the effects of the earth's magnetic field upon the rays, which also indicated a positive rather than negative charge. "There is no indication whatsoever of negative rays," Johnson reported. "Measurements are complete enough to leave little room for doubt."

If Johnson was correct, it appeared that both Millikan and Compton were both partly right and partly wrong. "Perhaps these measures will help to finish up the scientific picture of cosmic rays," said Johnson. "Perhaps we have arrived near the solution of the enigma of these rays." At least the *Philadelphia Inquirer* seemed to think so, headlining their report "Both Sides Declared Right in Cosmic Ray Arguments." Even the always pro-Millikan *Los Angeles Times* was hopeful for a peace agreement, stating that Johnson's findings "should go a long way toward bringing [Compton and Millikan's views] to the point of acquiescence."[19]

Perhaps so. One way or another, that was not going to stop the ascent of the enormous balloon in Chicago, where the mooring stakes had been pounded down into the ground of Soldier Field and Settle, checking weather maps at the Weather Bureau office, declared "It's O.K" to launch. The flight was on.

On Saturday, August 5, 1933, all the preparations and delays finally came to an end. Last-minute tests and preparations had delayed the planned 1:00 a.m. takeoff, but nobody seemed to mind, caught up in the spectacle. At least twenty-five thousand spectators in Soldier Field had been watching the pre-launch ceremonies, including parades, band concerts, a brief speech by Settle, and all the various preparations, including the inflation of the balloon gas bag with hydrogen. Jean Piccard stood by proudly, helping to supervise the proceedings and signing autographs. Arthur Compton was away in New York but present in voice, wishing Settle good luck over a loudspeaker hookup from the East.

At last, at 3:00 a.m., the crowd's patience was rewarded as *A Century of Progress* lifted off into the darkness, illuminated by three army floodlights. Outside the stadium, at least twice as many more spectators also watched. As the balloon rose and a band played the national anthem, Settle waved to the cheering multitudes and disappeared inside the gondola, sealing the hatch behind him.

The balloon swiftly climbed straight up to almost a mile and then began to veer off slowly to the west, out of reach of the searchlights. This was not the way things were supposed to go. Outside Soldier Field, the crowd started to follow the moonlit shape of the balloon as it continued westward and then began to fall.

14.2 The *Century of Progress* balloon undergoes inflation at Soldier Field. (Wikimedia, U.S. Naval Historical Center)

Ten minutes after its majestic ascent, *A Century of Progress* came down in the middle of the Burlington & Quincy railroad yards. The gondola bounced once and then settled down on the tracks, the quickly-emptying gas bag falling all around. Settle quickly clambered out of the slightly dented craft and pulled a ripcord to release the remaining hydrogen, shouting at the rapidly assembling spectators to put out their cigarettes. Police, railroad men, and a team of Marines moved quickly to get the crowd under control and keep them from trampling and damaging the fabric of the now sadly deflated balloon.

The not-quite-as-deflated Settle explained the reasons for his premature landing to the press. Because the cooler nighttime air would prevent the hydrogen gas from expanding and the balloon from reaching full buoyancy, he had planned to hold overnight at about 5,000 feet until sunrise and the warmth of the day would allow a further ascent.

"I opened the valves to let out enough gas to hold me at that level," he said. "But once opened, I found I could not shut off the flow." During all the pre-launch festivities and preparations, the workings of the valve had been demonstrated by opening and closing it while spectators listened to hear the escaping gas. "Commander Settle later admitted that he was not quite positive the valve was completely closed as he took off, but was unwilling to spoil the show with further delay," reported *TIME* magazine.

Ever the dedicated aviator, Settle was undaunted. "I want to try this flight over again, but with a change in the valve system," he said. Even while the balloon was being gathered up and the gondola loaded onto a flatbed truck to haul it off for repairs, the sponsors of the flight, including the World's Fair, *Chicago Daily News*, and National Broadcasting Company declared that they and *A Century of Progress* would be taking another crack at the stratosphere as soon as possible. *TIME* noted that in case anyone had doubts, the *Chicago Daily News* coverage managed to spin the mishap into an unexpected triumph, calling it "a balloon flight that ended as brilliantly as it began," featuring "a thing that will be imperishable in the history of ballooning. [Settle] landed the biggest balloon ever built in the heart of one of the biggest cities in the world in darkness, and landed it perfectly."[20]

Settle had indeed pulled off a brilliant landing under difficult conditions and brought down Compton and Millikan's instruments undamaged in the process. But cosmic ray observations in Chicago railroad switching yards were of limited scientific value. Two days after the abortive flight, Millikan wrote to Compton about his grounded electroscope: "I see by the morning paper that it is not intended to attempt the flight again before October . . . it would be desirable from our point of view that the instrument be returned, first, because our program can use it very profitably with the next two months, and second, because we want to be quite sure that it is in condition for successful recording in the flight if it occurs in October."[21]

About a week later, two of Millikan's electroscopes departed on their solo ocean cruises, one aboard the Grace Line freighter *Charcas* to Chile and the other on a round-the-world odyssey aboard the Dollar Lines ship *President Garfield*.[22] "You must appreciate, of course, that in giving assistance as we can in this connection, the Company accepts no responsibility whatsoever as to the care, custody, or value of the instruments, and if the instruments are particularly valuable that you should take out insurance thereon before the vessel departs," the vice president of the Grace Line informed Millikan, adding that any such insurance should also make sure to waive "entirely rights of subrogation against the steamship company in case anything whatsoever should happen to the instruments."[23] Science and cosmic ray research were all very well, but the company must of course be protected at all costs.

With any hopes of settling the Millikan-Compton debate in Chicago now spread all over a West Chicago railyard in the form of an acre of rubberized fabric and wisps of hydrogen gas, matters seemed at an impasse. But more balloons were about to rise, more instruments carried to exotic environments, more data to be captured—enough to build a rising and compelling consensus on cosmic rays, whether Robert Millikan liked it or not.

15

RISING AND FALLING

Still in New York, Compton had already been composing a letter to Millikan when he received Millikan's request to return the electroscope slated for the World's Fair flight. Under the auspices of the Carnegie Institution and along with his frequent collaborator R. D. Bennett of MIT, Compton had been busy working on the design of a new "cosmic ray meter" that could be set up in a network of widely scattered stations to automatically provide a continuous record of data. The Compton-Bennett instrument was an attempt to build upon and enhance what Compton termed the "elegant simplicity" of the Millikan electroscope.

Although Compton admitted that their present design lacked that attribute, "we hope, however, that in the present design the loss of simplicity has resulted in the gain of correspondingly increased precision." Compton wanted to know whether Millikan might be willing to provide a recording camera as used on the Millikan electroscope, and also invited his comments on the new Compton-Bennett design. The Chicago flight may have failed, but it seemed that the new spirit of apparent amiable cooperation that it had engendered between the two men lived on. No record of Millikan's response, however, could be located in his papers.

On the subject of Chicago, Compton mentioned in a postscript that "we are naturally much chagrined over the fiasco of the flight, especially after several of our men have spent solid months of work preparing the apparatus." Although he wasn't aware when or if a second flight attempt would be made, he would return Millikan's instrument, while "hoping you will loan it to us again if there is a 'next time.'"[1]

The "next time" was not long in coming. Aside from the determination of Settle, the Piccards, and the World's Fair officials to move past the humiliation of the Chicago mishap, new motivation had appeared from an unexpected direction. Settle may not have broken Piccard's altitude record in Chicago, but a fresh challenger had joined the fray: the USSR. On September 30, 1933, three Soviet Air Forces officers took a hydrogen balloon to 11.8 miles, claiming a new world's record—at least unofficially. U.S. Navy Bureau of Aeronautics chief Rear Admiral Ernest J. King pointed out that although the Russian flight was "a very marvelous performance," any record claim would first have to be verified by the official international body governing such matters, the Fédération Aéronautique Internationale.[2] Soviet publicity made a point of emphasizing the flight's scientific objectives to "obtain more information regarding the cosmic rays."[3]

For Settle, the U.S. Navy, and the other backers of the Chicago flight, more information about the cosmic rays likely held far less importance than the thought of Communist aeronauts holding world records. Preparations accelerated for a new flight of the repaired *Century of Progress* in the fall that would unquestionably claim the altitude record for the United States, again lifting off from the fair in Chicago. Talk of world records was still being downplayed, however. Said the *Chicago Tribune*, "[Settle] has no hopes of setting a new altitude record. . . . The principal object of the trip . . . is to study the cosmic rays."[4]

Millikan's electroscope was still sitting in Chicago, "in complete readiness for the flight at a moment's notice," as Compton informed Millikan. Meanwhile, however, Compton had another idea. One of his new recording cosmic ray instruments was about to be given an ocean ride to New Zealand in the company of Professor P. W. Burbidge of Auckland University. Would Millikan like one of his own instruments to go along on the trip?[5] Millikan responded that there seemed little point, especially since he already had two of his electroscopes cruising the high seas at the moment.[6] Compton had apparently missed the various newspaper articles marking their journeys.[7]

Settle was ready to go again at the beginning of November, but the Chicago weather was not cooperating, so the entire operation was relocated to the Akron, Ohio, airport. Finally, at about 9:30 a.m. on November 20, Settle and his observer Major Chester L. Fordney of the Marine Corps[8] lifted off in *A Century of Progress*, carrying both Millikan's and Compton's instruments that were supposed to go up three months earlier.

After an eventful flight during which they lost radio contact with the ground and were briefly feared missing, Settle and Fordney finally turned up in the marshlands of southern New Jersey. Their onboard instruments confirmed them to have reached an altitude of about 61,000 feet, handily beating both Piccard and the unofficial Soviet mark. At least for the moment, America had its world's record, and Settle and Fordney were feted as America's new heroes of the stratosphere.[9]

In the midst of all the celebration of record breaking, however, some continued to insist on the more noble scientific aims of the feat. "Of course, their trip had a more utilitarian purpose than merely to give a thrill to venturesome men," opined the *Boston Globe*. "They were seeking data on the cosmic ray, that mysterious force which scientists know of but little about."[10] The *Los Angeles Times* noted, "Those who have been following the various phases of the Millikan-Compton controversy will await the outcome with great interest."[11]

That outcome would have to wait until the scientific instruments were brought back to Chicago so the collected data could be analyzed. Results would not be long in coming. Only a couple of weeks later, Compton announced that his data from the Settle-Fordney flight showed that cosmic rays were a hundred times greater in the stratosphere than at sea level. "It is only a matter of time, Prof. Compton said, until the further tabulation and study of the experiments made in the balloon will sustain or disprove the different theories of the rays now held by Dr. Compton and Dr. R. A. Millikan of the California Institute of Technology," reported the *Chicago Tribune*.[12] The *Boston Globe* was more flippant, noting that if Compton's report was true, "this is probably an advantage of living at sea level."[13]

Any definitive word from Pasadena was yet to come, as Millikan seemed to prefer playing coy while he continued with his own airplane and sounding balloon measurements. At the next annual year's end

AAAS meeting, this time in Boston in the midst of a blizzard, both he and Compton refrained from any big pronouncements or heated debates.[14]

But a new wrinkle was added to the argument at the AAAS meeting by an outside party, Dr. J. Robert Oppenheimer. A brilliant theoretical physicist who divided his time between Caltech and the University of California at Berkeley, Oppenheimer presented some new ideas based on cloud chamber photographs revealing the formation of atomic particles from the disintegration of cosmic rays. Both Millikan and Compton were present for Oppenheimer's presentation, which had implications for the firmly held scientific conclusions of both men—photons/birth cries vs. charged particles. Again, it seemed, the theoreticians such as Oppenheimer were making life more complicated for the experimentalists such as Millikan and Compton.[15]

Just before leaving Chicago for the AAAS meeting, however, Compton sat down with a *Chicago Tribune* writer on Christmas Day for a lengthy "authorized" interview in which he pontificated on a matter that Millikan would no doubt have found quite close to his own heart: the spiritual and philosophical oneness of science and religion and the lack of conflict between the two. "Science can have no quarrel with a religion which postulates a God to whom men are as his children," he pronounced. "Not that science in any way shows such a relationship . . . but the evidence for an intelligent power working in the world which science offers does make such a postulate plausible.

"It is an inspiring setting in which we find ourselves," Compton continued. "As we recognize the greatness of the program of nature which is unfolding before us, we feel that we are part of a great enterprise in which some mighty intelligence is working out a hidden plan." He tied his own work specifically and the discoveries of science in general into a philosophy of cosmic purpose and meaning yet based on science. "His God is a working hypothesis, rather than a closed system of thought," explained the interviewer. "It is a God immanent in the world of nature. He is not afraid to test any new fact against it. . . . Dr. Compton's thinking leads him to choose a deistic view of the intelligence and order in the world."[16]

One might expect such sentiments to be expressed during the Christmas season, but for both Compton and Millikan, they were also quite sincere representations of deeply held beliefs. This was hardly the only time that either man had expressed such views publicly; both had done so quite frequently in speeches, press statements, interviews, and published lectures and books, offering some important insights into the intensity and the persistence of their ongoing public debate. Especially for Millikan but also to a profound extent for Compton, the cosmic ray questions they were arguing over and striving to solve weren't merely an interesting scientific problem. They were essential matters of God, the universe, the nature and purpose of creation itself.

Although not everyone thought cosmic rays quite so significant, they were obviously pretty important if Nobel Prize-winning scientists spent so much time chasing cosmic rays all over the globe and arguing about them. And because they remained a focus of controversy and unanswered questions, they were a source of awe, mystery, and wonder for nonscientists and the public at large, something upon which all sorts of hopes, dreams, anxieties, and fears could be projected and indulged. Much of this centered around two general yet related ideas: that cosmic rays would provide a new source of inexhaustible energy for mankind, or that they would become the basis of new weapons that could wipe out cities in a single blow. Or perhaps cosmic rays were both, a new form of Promethean fire that could build or destroy depending on the wisdom or folly of man's own choices. Such notions presaged the popular ideas about atomic energy, then still only a fantasy of comic books and science fiction, that would become pervasive after Hiroshima and the coming of the Cold War.

In some ways, it was only natural that such wild ideas would pop up in popular discussion, when almost every fresh idea, speculation, and opinion from acknowledged experts such as Millikan and Compton was so keenly followed and reported and extrapolated upon, even before data could be fully analyzed and experiments completed. For the average member of the public, it wasn't necessarily much of a stretch to go from eminent scientists talking about the birth cries or heat death of the universe, invisible super-penetrating rays from outer space of immense energy, and new subatomic particles appearing out of the void, to believing in the feasibility of super cosmic death-ray weapons or new energy sources that would close all the coal mines, propel ships and airplanes and automobiles, and heal all known diseases. And while Millikan, Compton, and all the other noted scientists working on cosmic rays would consistently dismiss such ideas whenever asked about them, they were also far too busy with actual scientific work to dedicate too much time in any efforts to dispel such widespread and persistent misconceptions—leaving the field wide open to crackpots, con artists, fabulists, and other dedicated purveyors of nonsense.

It also didn't help when nearly every new announcement from Millikan, Compton, and their colleagues was so often treated as the final answer that would settle all the questions at last, such as Compton's announcement of what the Associated Press called an "amazing atomic mist" in the stratosphere. Found by Compton's instruments aboard the Settle-Fordney flight, this "mist" was apparently composed of "flying particles of subatomic size which may be older than the earth itself—perhaps bits of the original creation." As Compton's paper on the results describes, the findings seemed to indicate either a new variety of cosmic ray or subatomic collision fragments consisting of protons, positrons, or other particles. As usual, the *Los Angeles Times* was not convinced: "Just how this Compton 'discovery' can in any way upset the Millikan theory it is difficult for a lay student of the subject to see. To such it seems more likely to confirm it." Meanwhile the *Philadelphia Inquirer* rhapsodized: "It would seem that beyond our atmosphere there are deathless energies, older than the earth, perhaps as old as time. That man does not understand them now is not to say he will never read their meaning."[17]

Notwithstanding the ever-staunch support that Millikan enjoyed from the *Los Angeles Times*, the evidence for the particle nature of cosmic rays was continuing to steadily build, even as he continued to steadily deny it. Finally announcing his own results from the Settle-Fordney flight, he insisted that they supported his original hypothesis of atom building in deep space. "His conclusions still differ fundamentally in several important respects from those reached by Dr. Arthur H. Compton," observed William Laurence, including that the rays were largely photons, not particles. "Practically all cosmic ray ionization is due to the passage of positive and negative electrons, rather than protons, alpha-particles, or heavier nuclei, through the atmosphere," Millikan told the American Physical Society. Those electrons provided no more than 3 or 4 percent of the ionization seen at sea level, but they were responsible for the geographical variations observed, such as the latitude effect. Otherwise, said Millikan, "the greater part of the ionization of our atmosphere is due to photons."[18]

So the debate continued. As both he and Compton continued to gather further data and argue, the next major manned stratospheric balloon extravaganzas were getting ready to take flight, in another bid to break records and to read the meaning of the cosmic rays.

Following their unofficially acknowledged record flight the previous autumn, the Soviets undertook another high-altitude adventure in January 1934, but this one ended in tragedy. The three-man crew of the *Osoaviakhim-1* were killed when their craft went into an uncontrolled descent and crashed.

The Army Air Corps and the National Geographic Society were not going to let that happen with their mission.[19] At the end of December, Captain Stevens, who would go along as scientific observer along with two other crew, had filled in Millikan on developments, including funding arrangements and the construction of the gondola, dubbed *Explorer*. Although Millikan was decidedly on board, it was unclear whether Compton would join the effort this time, even though the nine-foot-diameter magnesium alloy gondola would feature plenty of room for scientific instruments.

"Whether Dr. Compton will be interested in our flight is something that we do not know," Stevens wrote. "Last summer Mr. C. F. Kettering approached him, and Dr. Compton told him that he was not interested, unless the flight be made in other latitude than this. . . . Let us hope that Dr. Compton may change his mind."[20]

Compton would not change his mind, apparently quite content with pursuing his own research program and seeing little point in repeating the same sort of observations already made on the Settle-Fordney flight. Instead, Millikan's instruments would have the company of the Bartol Research Foundation, home of dedicated cosmic ray chasers W. F. G. Swann and Thomas Johnson.[21]

As the Army's plans continued to progress for a flight sometime in July, other parties were putting together their own excursions, all of which would prominently feature cosmic ray experiments. In preparation for the Piccards' next sky voyage in Canada, Jeannette Piccard, wife of Jean, was finishing up work on her balloon license, becoming the first woman in the United States to earn such a credential.[22] Admiral Byrd was already in Antarctica on his second expedition, along with Compton's instruments.[23]

On July 28, 1934, Captain Stevens, Major William Kepner, and Captain Orvil Anderson lifted off in the *Explorer* underneath the largest free balloon yet constructed from the "Stratobowl," a natural bowl-like depression in the remote Black Hills of South Dakota. At first all seemed to be going well, then began deteriorating quickly. As the balloon attained an altitude of just over 60,000 feet, the huge gas bag began to tear open, hydrogen escaping from multiple gaping holes fifty and sixty feet wide, and the balloon began dropping rapidly. The crew had to bail out by parachute, as the gondola crashed to earth in rural southern Nebraska.

Not only were no records set, but it also appeared that the scientific objectives of the flight had been lost, most of the equipment destroyed in the totaled gondola. A roll of film from one of the Bartol Research Foundation instruments was later salvaged, but little else. About four feet of film from one of Millikan's three onboard electroscopes was also recovered. After all the publicity and hype, the failure of the mission was something of a scandal for the Air Corps and the National Geographic Society, especially so soon after the Soviet disaster in January, and official investigations followed. But Stevens soon began lobbying for another attempt in a redesigned balloon and gondola.[24]

Also undeterred were the Piccards, who announced that their next balloon flight would rise from Detroit in October. There would be no attempts to set any records; the sole purpose was to study cosmic rays.[25] Jean and Jeannette Piccard launched from Ford Airport on October 23, 1934, carrying a Millikan electroscope and other instruments from Bartol's W. F. G. Swann, though none of Compton's. Unfortunately, their flight would also end prematurely, though not nearly as spectacularly as the army flight. After reaching a height of about ten miles, they began to run into visibility problems and came down for a slightly rough landing in a treetop on a farm near Cadiz, Ontario. Jean got away with a sprained ankle and Jeannette with a few scratches, but otherwise they and the scientific instruments escaped intact.[26]

1934, it seemed, was not the year for piercing the stratosphere, at least not for human beings. It was also beginning to appear rather unnecessary, at least as far as cosmic ray research was concerned. While the army and the Piccards had been trying to carry instruments to new heights, Arthur Compton had been busy developing a means to do the same thing without any human involvement at all.

Working with J. M. Benade of Forman Christian College in Lahore, India, whom Compton had befriended on his previous Indian visit some years before and who was now in Chicago on sabbatical, Compton began testing small "flying laboratories with radio brains," otherwise known as radiosondes, from the University of Chicago campus, which would automatically transmit data from the sky to a ticker tape recorder on the ground. "One important feature of the plan, Prof. Compton pointed out, is that the balloon and its instruments may be lost forever without harm, inasmuch as its messages will be a matter of record on the ground," noted the *Chicago Tribune*.[27]

It seemed Compton had learned a lesson from Millikan's lost balloons in Texas and elsewhere. Reward money would still be offered for a returned instrument package, but its total cost was only about $100, so any that were lost could be considered acceptable losses—the all-important data would still be collected. Compton allowed that although this method was not expected to replace measurements made from large stratospheric balloons that could carry heavier equipment, "much information, however, can be obtained with these lighter balloons," he said. "This will supplement, in regions difficult of access, the more detailed information which the larger balloons give us in ordinary latitudes."[28]

After several weather-related delays, Compton and Benade lofted a fifteen-foot diameter hydrogen balloon carrying a ten-pound cardboard-encased package containing shortwave radio, barometer, and thermometer from the roof of the Ryerson Laboratory on August 29, 1934. It was only supposed to be a brief test of the concept to a modest 1,000-foot height with the balloon tethered to the roof, but a gust of wind broke it free and it soared into the Chicago evening sky. It rose almost eighteen miles, faithfully transmitting data over the shortwave radio until the balloon finally burst and all fell to earth about thirty miles southeast of Chicago and was lost.[29]

Still, Compton and Benade considered the unplanned test flight a complete success, fully proving the workability of the concept. And since

15.1 Weary of the vagaries of manned balloon flights, Compton switches to placing his cosmic ray instruments aboard radiosondes. (University of Chicago Photographic Archive, [apf1–01858], Hanna Holborn Gray Special Collections Research Center, University of Chicago Library)

no cosmic ray instrument had been aboard this preliminary test, no great harm was done in any case. "The test showed the adequacy of our barometer and of the radio method of noting its readings," said Compton.[30]

Along with his program to establish permanent monitoring stations in strategic geographical locations, it would comprise a major element of his new cosmic ray campaign supported by the Carnegie Institution, which was becoming somewhat more circumspect and conservative in doling out funds to the various separate cosmic ray programs it had been bankrolling, which had begun to seem redundant and exorbitant to some Carnegie officials. Compton was also still wrapping up his ambitious world survey, of which the Byrd expedition remained the major uncompleted piece.

As for Millikan, he was proceeding doggedly with his own experiments, including enlisting his collaborator Serge Korff to undertake a series of high-altitude airplane flights in Peru to collect more equatorial data. Much to the surprise of his scientific opponents, however, he was about to announce a change in his fervently-held positions—but only to a very slight, practically inconsequential extent. He would appear to give up something, while claiming something new that he argued still ultimately vindicated him.

16

CREATION AND ANNIHILATION

Whether in dress, personality, politics, religion, behavior, or his professional life, Robert Millikan was staunchly and proudly conservative, the farthest thing from a radical thinker. He liked stability, he liked order, he liked everything in its right place and everything functioning in predictable, stable, logical ways. He was methodical, painstaking, systematic, thorough in everything, not only in his personal life but especially in his professional life.

A byproduct, or perhaps a complement, to all those qualities was what some might call a firm, unwavering strength of conviction and others might call an intransigent, unyielding stubbornness. He was not a man to change his mind easily or to be swayed by shifting popular opinion. He was, in the words of Marcus Aurelius, "like the rock that the waves keep crashing over. It stands unmoved and the raging of the seafall is still around it."[1] That was Robert Millikan, although he would undoubtedly have stated the same sentiment in some biblical turn of phrase rather than a quote from a Roman emperor.

That trait especially came into play wherever his fundamental philosophical, religious, and spiritual convictions were concerned, as with his ideas of universal birth cries and atom building. The concept of an eternal, unchanging universe in which his Creator is still and always on the job meshed perfectly with the cosmic ray theories that he had developed and set in his mind not long after he had begun investigating the subject before World War I. It formed the bedrock of all his subsequent work, with each new cosmic ray finding, whether by himself or others, apparently reinforcing his favored cosmogony: the confirmation of an extraterrestrial source, the three-band energy level model as described in chapter 4, the photonic nature. The ideas were so appealing, so harmonious, so obviously *right* that anything that seemed to contradict them could be dismissed, tossed aside, or explained away and disposed of, nothing more than a bit of momentary static in a radio broadcast of a beautiful symphony.

Arthur Compton had posed the first major threat to Millikan's beautiful music, a challenge that couldn't simply be ignored or disregarded. And as the continuing evidence presented by Compton and others rose in a relentless and ever-strengthening floodtide, Millikan found himself struggling to hold it back before it completely swept away all his cherished philosophies. He had to find some way of both giving in to the inevitable yet also hanging on to whatever he still could, to go with the flow somehow without being swept away by it.

Any good scientist, of course, needs to be able to change their mind whenever the preponderance of evidence demands it, no matter how appealing an old, established idea may be. And Millikan, however conservative, had certainly done so more than once. It took him quite a while to fully accept the revolution in physics brought by Einstein and quantum mechanics, but he had finally done it. Originally convinced that cosmic rays were of earthly origin, he changed his mind there as well, so much so that he had even coined the term and was still often believed to have discovered them. Stubborn he was, but he was also a scientist.

There was a limit, however. Now the inexorable weight of scientific consensus was pressing his back against the wall of his most profound spiritual nature and, not incidentally, his healthy professional ego. If he was going to acquiesce at all, he would find a means to do it his way.

At the October 1934 International Conference on Physics in London, Millikan gave a presentation that both startled and annoyed his colleagues. In what was termed a "modification" of his theories, he announced that he had concluded that cosmic rays were not only evidence of continuous creation, but also annihilation as well. He could no longer deny the existence of a latitude effect, nor that some cosmic rays had been confirmed at energies far above those he had earlier declared possible because they were inconsistent with the atom-building hypothesis. Those extremely high energies—as much as 10 billion electron volts detected at equatorial latitudes—could only be explained by the total annihilation of atoms in space. The lower energy rays, the ones affected by the earth's magnetic field and thus manifesting the latitude effect, were resulted from the creation of more complex atoms from hydrogen, that is, the good old atom-building hypothesis. In both cases, however, they consisted almost entirely of photons, not charged particles.[2]

To the *Los Angeles Times*, it only seemed that Millikan had provided "material for further argument by mathematicians and astronomers . . . Dr. Millikan appears to be meeting [Prof. James] Jeans halfway. It will be interesting to hear what the latter has to say about this revised theory."[3]

Others were less certain. "Those who have followed the researches of Professor Millikan were astonished to learn that . . . he had declared that the cosmic rays originate in the destruction of matter," wrote Waldemar Kaempffert of the *New York Times*. "It looked as if he had repudiated his long-maintained view."[4]

But that was only a matter of appearance. To Millikan's assembled colleagues, it seemed as though he were almost trying to appease his critics by throwing them a bone, trying to have it both ways while still hanging on to his increasingly untenable position. Compton wrote to his colleague R. D. Bennett that the conference "was largely a stage for Millikan to present his views, which were received with a distaste approaching nausea by his British listeners."[5]

Millikan responded to the distaste by moving to explain himself further in more public venues. A lengthy article in the *New York Times* summarizing his "new" theory, as well as a published interview with Science Service correspondent Donald Caley, are almost equal parts scientific explication and defensive self-justification, as if he were offering up alibis for any supposed errors he had previously made. Phrases such as "I have been repeatedly pointing out . . . I myself have never used that phrase . . . I am only reiterating the view . . . I have seen no reason in the past for assuming . . . this is what we all agree produces the so-called 'latitude effect' . . . I was merely pointing out . . . as I have repeatedly stated . . ."

have the effect of making Millikan's offerings seem to be more of a defiant mea culpa or an elaborate exercise in excuse making.[6]

He would continue similarly at the next American Association for the Advancement of Science annual meeting, this time in Pittsburgh, where the highlights would include the attendance and lecture of Albert Einstein and the election of Arthur Compton as the next AAAS president. Among the public exhibits would be one provided by the Mellon Institute that would demonstrate "that enigmatic toy of modern science—the cosmic ray."[7] Although Einstein would tell reporters at the meeting that recent talk of atomic energy from the nucleus of the atom was unlikely—"it is like hunting birds in a country where there are very few birds, and in pitch darkness"—the cosmic ray, at least, was still a big draw.[8]

In Pittsburgh, however, probably because he was speaking not before an audience composed wholly of scientists but a committee of science teachers, Millikan offered not a mea culpa but a sermon, laying down not a series of self-justifications but more of a set of proclamations from the Mount, playing the role not of scientific savant but of an old-fashioned minister. That was evident from the title of the talk, "What to Believe About Cosmic Rays." He set out eight "articles of faith" that constituted the present "cosmic ray credo." Reverend Millikan's "credo" covered the questions that were essentially settled, such as the extraterrestrial nature of the rays and their extreme penetrating power, and others on which "the jury is still working," such as the processes that created cosmic rays and whether they were photons, particles, or both. On that latter question, he was confident that the jury will "hand in its verdict within a twelvemonth." That certainly sounded optimistic, but as William Laurence observed, Millikan's "cosmic ray credo" excluded "some of the most important hypotheses on both sides of the controversy, including those of his own school."

In fairness, given his audience, Millikan was likely consciously adopting a somewhat tongue-in-cheek tone, despite his admonitions to the teachers including "do not report to your pupil any conclusions as even probable until two or three independent observers get into agreement on them." He also seemed to sympathize with the "hopeless confusion" the public suffered in considering the whole subject. "It is just too bad to drag an interested public through all our mistakes, as we cosmic ray experimenters have done during the past four years." He could also not resist another shot at the press: "This situation is not improved by the existence of the daily newspaper, which, as its very name implies, is under a greater pressure to find for its pages something that is new rather than something that is true."

But he was also striving to depict the subject as far less contentious and closer to a final settlement than usually depicted—a final settlement that would undoubtedly vindicate his own ideas which he had just set out. Facts were established in physics, according to Millikan, when "nine-tenths of the informed and competent workers in the field are in agreement upon it." Of course, there would always be "a small percentage of people who will vote 'no' . . . here one has left the field of physics and entered the domain of psychology or perhaps pathology."

The last point of Millikan's "credo" noted that "you may, of course, believe any direct experimental findings from which the personal equation and the judgment of the observer have been entirely eliminated," a rather ironic statement coming from a scientist who had demonstrated himself to take the entire subject very personally indeed.[9]

While Millikan was certainly correct in thinking that most of the press continued to relish the spectacle of two Nobel Prize-winning scientists publicly bickering, there were some exceptions who were perfectly willing to depict the entire situation in Millikan's terms, as no more than a friendly disagreement or minor misunderstanding. The *Los Angeles Times* had been taking that line fairly consistently, as if to imply that Compton and the other misguided souls who failed to acknowledge the preeminent wisdom of the "Caltech savant" would sooner or later see the error of their ways.

Commenting on Millikan's latest speech and his remarks about the press, the paper's science columnist William S. Barton noted, "Dr. Millikan appears to have been referring in part to the excitement and publicity occasioned by the conflicting theories of himself and Dr. Arthur H. Compton. . . . Again, Dr. Millikan and Dr. Compton were portrayed as being at swords' points over the nature of cosmic radiation." But that was only a "misunderstanding," and "what may prove to be the decisive test will be made this year by Dr. Millikan," who was planning to conduct balloon flights equipped with radios to transmit data to the ground, which "will obviate the previous difficulty due to loss of recording instruments." No mention was made of the fact that Compton had already been doing the very same thing for at least several months.[10] In fact, the balloon flights Millikan would begin that summer would feature the same arrangement as the various others he had been conducting for years: carrying an instrument package with a $1 bill inside and a note promising $5 more for safe return of the payload.

Compton was spending much of the year in England as a guest instructor at Oxford, but he returned home for some time in the spring to present a few more lectures, send up some more automated radio balloons, and check over the data received from the latest Byrd expedition, which had just returned from Antarctica. The South Pole data again confirmed a sharp rise in cosmic ray intensity compared to the equatorial regions, adding to the ever-mounting pile of evidence that they were particles. "We know that the cosmic rays are affected by the earth's magnetic field and hence are in part electrical," he said. He was also becoming more convinced that they were of galactic or perhaps extragalactic origin.[11]

By this time in the decade of the 1930s, the cosmic ray phenomenon had developed two distinct yet related personalities. There was the serious, professional, scientific side, as evidenced by the ongoing Millikan-Compton debate, the various experiments and expeditions, the announcements of new discoveries and theories. Then there was the public side, which was decidedly less serious and considerably sillier. Along with all the fodder for snarky editorial writers, comic books, and movie serials, cosmic rays had just the right amount of mystery and otherworldly spookiness to provide golden opportunities for less savory individuals, such as self-styled "healers," crackpot inventors, and quasireligious groups such as the Rosicrucians. In Los Angeles, several "cosmic ray health centers" would cause trouble for the City Council. A horse named "Cosmic Ray" would even become a brief fixture on the racing circuit.

Robert Millikan, Arthur Compton, and their colleagues might still be hard at work trying to pin down the precise characteristics and origins of cosmic rays, but there were apparently many far less learned individuals who already knew not only exactly what they were but also how to use them for the betterment of humanity—or at least line their personal pocketbooks.

Back in the realm of science and scientists, Millikan again raised some eyebrows and caused some consternation with the publication of a revised and expanded edition of his 1917 book *The Electron*. While that volume dealt mainly with his lengthy experiments to determine the charge of the electron, this one covered much more, its new title reflecting the many developments that had occurred since its original publication almost twenty years before: *Electrons (+ and –), Protons, Photons, Neutrons, and Cosmic Rays*.

Some reviewers acclaimed it as yet another triumph of the eminent Dr. Millikan, such as reviewer Philip Kinsley: "It is not an easy book to read and much of it is beyond the layman, but there are bits here and there which may enlarge one's understanding of the great search of science for the 'why' of all things. The nature of cosmic rays, the dogma of the heat death of the universe, which has puzzled so many scientists as well as theologians, are given careful treatment."[12]

Some readers who were not laymen, however, took a more jaundiced view, such as *Nature* reviewer L. G. H. Huxley, who especially singled out the chapter on cosmic rays, which appeared "to be directed towards establishing the hypothesis that the primary radiation is composed principally of photons, a view of its nature not generally accepted. Consequently other investigators are treated quite summarily."[13] Edward Condon in the *Review of Scientific Instruments* also thought that Millikan spent too much time emphasizing the work of himself and his collaborators while giving short shrift to the work of others, suggesting that perhaps the book should carry the subtitle of "Happy Days and Nights in the Norman Bridge Laboratory."[14] Millikan biographer Robert Kargon observed, "Millikan's stubborn refusal to agree with what was emerging as a consensus on cosmic ray issues, and his single-minded concentration on his own contributions and those of his collaborators, elicited criticism from many quarters."[15]

While both Millikan and Compton continued launching their unmanned balloons, the army and National Geographic Society were preparing their next manned stratospheric venture, hoping to overshadow the ignominy of the previous year's failure. This time it would be Captain Stevens and Orvil Anderson in a new and improved gondola named *Explorer II*, under a balloon filled not with hydrogen but with helium for added safety. Again launching from the South Dakota Stratobowl on November 11, Stevens and Anderson achieved an unofficial altitude of about 74,000 feet, later revised to a new official world's record of just under 73,000.[16] They came down in eastern South Dakota as America's newest heroes, feted by President Roosevelt in Washington.[17] Their record would stand for almost twenty years.

Neither Millikan nor Compton participated in this flight, although it also carried the by-now-obligatory cosmic ray instruments to provide some scientific legitimacy to the aerial record setting, built by the Bartol Research Foundation's W. F. G. Swann. Compton, at least, was reportedly "very interested" in learning what data had been collected by Swann's equipment, given the extreme altitudes attained by Stevens and Anderson and their in-flight reports of extremely high cosmic ray counts at those heights. "That represents a very important advance over previous work," he pointed out. "We engaged in the research are looking forward with enthusiasm and hope to find just what was accomplished."[18]

But as far as the American public was concerned, it was the record that represented the main objective and the chief triumph of the flight, not the scientific observations that Stevens had made on the way. As aviation historian Craig Ryan pointed out, "The two men [Stevens and Anderson] were lionized—but not, certainly, as pioneering researchers. They were celebrated, instead, as intrepid explorers. They were sanctified heroes at a time when America needed heroes." Indeed, although "both the *Century of Progress* and *Explorer* gondolas were crammed full of scientific equipment and carefully prepared experiments . . . neither in the end made any earth-shattering contributions to scientific understanding. The truth of the matter is that none of the experiments carried aloft by stratospheric flights of the 1930s required the presence of a human being. If science had really been the raison d'être of these flights, unmanned balloons would have made more sense."[19]

After the mishaps and complications of the manned flights of the past several years, Compton had certainly already come to that conclusion. He might still be very interested in whatever data Swann had collected on the Stevens-Anderson mission, but his new preference in favor of unmanned balloons increasingly was being vindicated by the development and ongoing success of his radio-equipped sounding balloons. Millikan was deciding likewise, especially as he also began to launch balloons capable of transmitting real-time data to the ground, instead of relying on recording instruments that depended on safe recovery (or their intact return by private citizens who found them) after returning to earth. Foreshadowing the attitudes of James Van Allen, one of the next generation of cosmic ray researchers who was an outspoken critic of the scientific value of manned spaceflight over unmanned probes, Millikan soon went beyond merely declining to participate in any further manned excursions. "Millikan eventually became a vocal opponent of scientific research funds being used on manned balloon flights, charging that the scientific agenda for *Explorer* was a 'sham,' and that manned balloon activities in general were a passing 'fad.'"[20] Van Allen became famous (or infamous) for many similar, almost word-for-word comments thirty years later as the Mercury, Gemini, and Apollo programs were grabbing headlines and congressional largesse.

Although cosmic rays continued to feature in the public consciousness as the 1930s began to slide toward the 1940s, with headlines announcing the solution to the cosmic ray mystery or the end of the Millikan-Compton feud—or both—appearing with a reassuring if unconvincing regularity, other events were pushing forward into greater prominence. The Lindbergh kidnapping saga was reaching a head with the trial of Bruno Hauptmann. The Depression was dragging on. On the other side of the world, Europe was continuing to heat up with the ever-more-threatening moves of Hitler and Mussolini.

With events of such import claiming more and more space in the nation's newspapers and on the airwaves, the Millikan-Compton matter was shoved ever further onto the sidelines, mentioned only occasionally when some fresh results, some interesting technical developments, exotic trips, or significant journal articles were announced. Anyone who had been following the affair more or less casually could be forgiven for assuming that, for all intents and purposes, the whole controversy had blown over and been resolved, save perhaps for a few esoteric details like the placement of a decimal point or a term in an equation or two.

But such an impression would be far from correct. In fact, Robert Millikan and Arthur Compton were about to move into a new stage of the battle, now joined by another party with a serious stake in the outcome. Unlike the Atlantic City showdown or the other contretemps conducted in the full glaring view of the public and press, this new phase would occur completely behind the scenes, but would be no less contentious. In some ways, perhaps because of its offscreen, out-of-sight nature, it would be even more bitter and nasty than anything that had happened before.

17

QUARRELS IN THE FAMILY

One of the observers that seemed to be leaning toward the idea that the Millikan-Compton conflict was all but history was *TIME* magazine. Reporting on the latest annual AAAS meeting, this one in St. Louis, the periodical provided a handy summary of the entire affair, beginning at the beginning. And that is what Arthur Compton also did in his opening address, his first as incoming president of the association.

Compton sketched out the early history of cosmic ray studies from the beginning of the century up to the present controversies, laying out the basics, mentioning all the important waypoints of experiment and discovery (including Millikan's), leading up to what was rapidly becoming the accepted consensus: most cosmic rays were particles, some negative, most positive, with a small residual of photons. Compton's belief was that they originated from beyond the Milky Way galaxy. He smoothly laid out all the work that supported these conclusions, not only that of himself but of others. "For most of the audience," *TIME* reported, "it marked the end of the 'mystery' of cosmic rays, wrote finis to one of the most reverberating scientific controversies of the century."

Millikan had been among the audience. "He did not rise, when the speaker had finished, to challenge his conclusions or even to ask a question. Impartial observers were therefore ready to write off their classic controversy as closed, to call it a cosmic clearance. If Dr. Millikan still cherishes the conviction that most of the cosmic rays are photons, he stands almost alone. Three years ago [in Atlantic City] he remarked that if he ever wanted to change his mind, he hoped he would not be pilloried. He has not been pilloried." As if to prove the point, the TIME cover featured a smiling Compton holding one of his cosmic ray gizmos.[1]

It seemed to be a neat and satisfying ending to the entire thing, a tidy resolution almost worthy of a Hollywood potboiler. Unfortunately, *TIME* was being both premature and too optimistic. Dr. Millikan did indeed still cherish his convictions, but he was again being discreet and refusing to be drawn into what he considered to be an unseemly public disagreement. He was going to take matters to a more professional forum, out of the sight or scent of the ever-hungry newshounds.

Compton was undoubtedly far too perceptive and too familiar with the personality of his scientific rival to take seriously *TIME*'s rosy assessments of lasting peace at hand. He continued to amass more data supporting the latitude effect, the latest effort consisting of the installation of one of his cosmic ray meters on the steamship *Orangi* out of Vancouver,

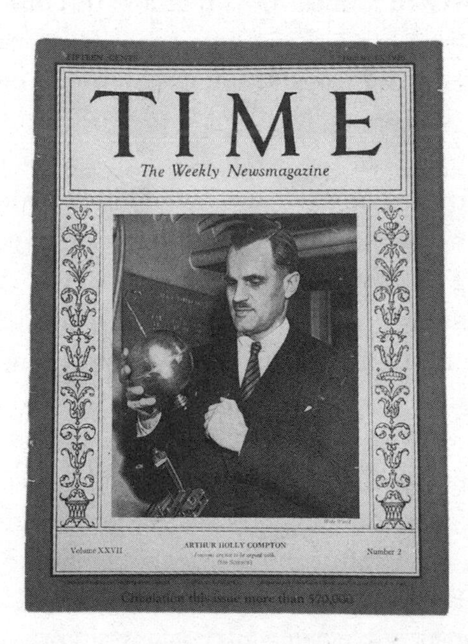

17.1 Compton makes the cover of *TIME* magazine in 1936 as the Millikan-Compton debate is prematurely declared over. (Wikimedia Commons)

Canada, en route to Sydney, Australia. Compton and his wife Betty would sail aboard the Canadian Australasian Line vessel and disembark on its stop in Honolulu, enjoying a nice Hawaiian vacation while the ship and Compton's instrument continued to Australia and back, picking them up again on the homeward voyage. Other duplicate instruments were also now operating at six other far-flung locations. Upon arriving back in Vancouver, Compton was pleased with the results of the cosmic ray meter's trip: "The latitude effects show up very clearly," he said. Presumably he and Betty were equally pleased with their brief tropical idyll.[2]

He also began work on a book about cosmic rays, which would be his first on the subject. It would be a comprehensive treatment, both historical and scientific, from the first inklings of the existence of a mysterious radiation to the current state of knowledge. From the surviving draft excerpts in Compton's papers, the book does not seem to have been intended as a full-fledged textbook for advanced students but more as a detailed but still accessible treatment for a scientifically astute but still general lay audience; it is not altogether free of equations and technical details but seems to lack the rigor found in a text for serious physics students. Compton apparently worked on the project sporadically for the next several years, enlisting colleagues Manuel Vallarta and later Bruno Rossi as collaborators and coauthors, but other obligations, distractions, and finally the major interruption of a new world war intruded. Compton and Rossi attempted to revive the project again in some form after the war, but finally decided to abandon it. "The advent of the war and subsequent occurrences beyond our control have made it impossible for us to complete the preparation of this book," Compton informed his contracted publisher McGraw-Hill.[3]

Compton's motivations for writing such a volume were no doubt serious and scholarly, but it's likely that he was also hoping to put on the record some kind of definitive statement and explanation of the consensus on cosmic rays that Millikan continued to obstinately resist. "We now know that these rays come from far outside the earth's atmosphere, and when observed at high altitudes are many times as intense near the magnetic poles as near the magnetic equator," he wrote in an early rough draft of the opening chapter. "This implies that they consist chiefly of electrically charged particles." Several paragraphs later he stated, "How they originate is still obscure; but increased knowledge of their characteristics has helped to limit the types of hypotheses that are admissable [sic]."[4] Had the book ever seen publication, one imagines that Robert Millikan would take issue with such statements.

Whatever *TIME* magazine might think, Millikan was not at all giving up the fight. The data collected by Serge Korff during his various equatorial adventures merely served to further confirm his views. "Dr. Robert A. Millikan disclosed tonight that he is standing by his guns in his four-year discussion with Dr. Arthur H. Compton . . . in the light of recent research he maintains his belief that at least 50 percent of cosmic rays consist of photons," reported the *Los Angeles Times*.[5] "Professor R. A. Millikan still insists that [cosmic rays] are bundles of light—photons," the *New York Times* noted.[6] And *Los Angeles Times* science reporter William Barton devoted an entire column to describing the state of things, complete with a photo of Millikan and his "friendly foe" Compton in which neither man looked very friendly. "The word friendly is used advisedly," Barton admitted, "the two greatest American physicists being on such close terms that they recently elected to travel together by train."[7]

Barton alluded to the recent *TIME* piece "which proclaimed Dr. Compton had scored a sweeping victory over Dr. Millikan. . . . Later data, however, give a vigorous new lease of life to Dr. Millikan's photon theory." That data essentially was the difference between cosmic ray intensity at high compared to low altitudes. Throughout his piece, Barton strove mightily to depict the debate in cordial terms. "Even honest Dr. Compton admits that whether he or Dr. Millikan's theory regarding the nature of the rays is correct should be unimportant so far as their respective scientific reputations are concerned. Just as two great artists at a convention of artists might argue about the relative importance of decoration and representation, the Nobel Laureates argue at their meetings about cosmic rays. Their arguments are fruitful, stimulating the imaginations to point out new and more difficult problems to solve. Reporters only half understand some of this shop talk, the result being that the public concludes the debate is more important than the thing debated."[8] That was certainly quite true.

Unlike much of the other coverage seen in the pages of the *Los Angeles Times*, Barton's presentation of the issues at hand does treat Compton and his ideas with reasonable fairness. Still, he concludes, "Dr. Millikan declines to comment directly on his scientific differences with Compton, but a physicist who isn't so reticent said: 'You could give the decision to Compton in every argument and Millikan would still be so far ahead he'd be out of sight!'"[9]

Judging from the unfinished manuscript drafts, if Compton's cosmic ray book had ever been finished and published, it would undoubtedly not have faced the same criticisms of a "single-minded concentration on his own contributions" that had been leveled at Millikan's recent book. Compton was quite scrupulous in giving credit where credit was due, even where Millikan was concerned. Millikan, however, seemed to have more of a problem in that department. When in the summer of 1936 he and Victor Neher published the results of their own worldwide cosmic ray surveys over the past several years in *Physical Review*, entitled "A Precision World Survey of Sea-Level Cosmic-Ray Intensities" and "The Equatorial Longitude Effect in Cosmic Rays," Compton finally decided he had had enough.

On November 12, 1936, Compton sent a brief letter to Millikan, along with a draft of a new paper that was both a response and a rebuttal to Millikan and Neher's latest papers. He told Millikan that "in order to clarify the position of other students of the geographic distribution of cosmic rays, it has seemed imperative to Professor Clay and myself to publish some statement in the *Physical Review* which would amplify the recent papers by yourself and Dr. Neher." Professor Jacob Clay was the Dutch physicist who had first reported a cosmic ray latitude effect as far back as 1927; although his results had been largely disregarded at the time, they had since been recognized as correct—at least, by most. Compton was giving Millikan a look at his proposed response out of professional courtesy: "Before these are submitted for publication, we should be glad if you would acquaint us with any errors of fact or interpretation, and especially we wish you would let us know if we have drawn unfair inferences from our reading of your papers."

On his cover letter, in a gesture of even greater professional courtesy, Compton added a handwritten postscript, offering his congratulations on the announcement of Carl Anderson's recent Nobel Prize in Physics for the discovery of the positron, which Anderson would share with Victor Hess, who was finally being recognized for his discovery of cosmic rays. Compton added that he had even written to the Nobel Committee recommending the award, and that "it might just as properly have been a joint award to Anderson and yourself!"[10]

Any warm and fuzzy feelings that Compton's congratulatory postscript may have inspired in Millikan, however, would have been promptly squashed as soon as he delved into the enclosed manuscript, beginning with its title: "'Precision' Surveys of Cosmic Ray Intensity," with the word "Precision" in sarcastic scare quotes.

What followed was an extensive twenty-nine pages of detailed analysis of the Millikan-Neher papers, starting with a historical review of the previous work on the geographical distribution of cosmic rays that had been done by Clay, Compton, and others, to make up for "the fact that [Millikan and Neher] have failed to refer to any except one of the many earlier studies of the same subject by other investigators." Because of that failure, readers unfamiliar with the cosmic ray field might "readily infer that the omission of references to earlier studies means that the authors consider the other work as of relatively negligible value and their own results as essentially new. . . . The fact emerges that the cosmic ray survey of Millikan and Neher is not unique with regard either to its precision or its extensiveness." And their conclusions had either already been established by other researchers or were unjustified.

One can imagine Millikan sitting in his office at the Norman Bridge Laboratory, reading Compton's paper and becoming increasingly incensed as he scribbled outraged notes in the margins, such as his response to the preceding: "Unworthy personal attack without any sort of justification since no claims were made other than those contained in the data." To Compton's statement, "The geographic study of cosmic rays was initiated by three trips made by Clay," Millikan remarked, "Completely incorrect as stated." To another statement that at least until the end of 1932 Millikan had been unwilling to acknowledge a latitude effect at sea level and had said as much in Atlantic City in 1932, Millikan scribbled: "This is grossly misleading. It is not a statement made at this time but a quotation from a much earlier article to show how I had stated my *earlier* failure to obtain a latitude effect." This was sophistry, because the published version of Millikan's Atlantic City address does contain the statement Compton describes, whether or not Millikan was quoting a "much earlier article."

Millikan made various other notations on his copy of Compton's draft, recording his initial reactions and perhaps also to serve as the basis of a more extensive response: "Not true . . . completely wrong historically . . . simply ridiculous!" He was likely also peeved by Compton's criticisms that failure to make appropriate corrections for the idiosyncrasies of Neher's electroscope, such as readings taken with and without the presence of lead shielding, had led to significant discrepancies in the data. The marvelously sensitive, vibration-proof, supremely accurate Millikan-Neher electroscope was really not the vaunted standard of reliability it had been advertised to be. "Because of these variations in the cosmic rays, more precise values of their geographic distribution can be obtained only by repeated measurements . . . many such data of increased precision are required before we can give an adequate answer to the questions now before us," concluded Compton.[11]

Before sending the draft of the paper to Millikan, Compton had also sent it to Jacob Clay for his comments and response, and to see whether Clay might want to add his name as coauthor. Clay responded that while he was "in complete agreement with you about the content and the tendency of it . . . I should feel ashamed to put my name under the head, as you have done the whole work." Instead, he suggested an additional statement of his own that would immediately follow Compton's paper in the *Physical Review*. Compton also sent Clay's response to Millikan along with his own paper.[12]

While Compton's article maintained a reasonably objective though critical tone, Clay's proposed addition was much shorter and far more pungent. "I fail to understand why Millikan in recent years has not followed the common practice in science of comparing his own results with those of other workers in the same field . . . if he had done this, he would have found that he has not obtained the precision reached recently by others. Due to this fact he takes a too simplified view of the problems at issue." Clay describes some of those problems, such as the complexity of defining the magnetic equator and its precise position and the extent of the influence of the earth's magnetism on cosmic ray measurements, questioning both Millikan's published results and his "attitude that he and his coworkers were the first to find this explanation" of the latitude effect. "We hope that Millikan and his collaborators will continue to cooperate in collecting cosmic ray survey data of such precision as will enable us to solve problems such as these."[13]

Clay was being far too optimistic. "My dear Arthur," Millikan wrote to Compton, "You have done the correct thing in sending me your article before submitting it for publication, but my commendation cannot go farther than this. It is so replete with misstatements, arising I hope from misunderstandings, and so incorrect in its attitude and mode of approach—argumentum ad hominum [sic] from beginning to end—that I think it will have a disastrous influence, if published, upon the esteem in which science and scientists are held by the public at large, and nothing whatever that I can do can prevent this."

It seemed to Millikan that if he simply kept quiet and ignored this "incorrect personal attack," he would preserve his own dignity and people would believe either him or Compton as they wished. But if he "corrected" Compton, the public "will simply look upon the whole performance as a dog-fight between two Nobel Prize men, and this doesn't help anybody." He would, however, write a personal response to the paper, although "it cannot be done quickly, as there is much too much to correct."[14] Millikan had not forgotten the press circus that had been surrounding the entire matter before, after, and especially during 1932 and culminating at the Atlantic City meeting.

Quite unexpectedly and unwillingly, the editor of *Physical Review*, John T. Tate, now found himself solidly caught in the middle, forced into the position of playing referee. Calling himself "very much perturbed," he wrote to Millikan noting that he had just received two papers from Compton and Clay about the recent Millikan-Neher papers, explaining that they presented two types of criticism, one about the interpretation of data and the other with "matters of priority and with your failure to refer to the work of others." He had no problem publishing the first type of criticism but was greatly reluctant to publish the second type.

Tate asked for Millikan's opinions on Compton and Clay's criticisms and how to handle the situation. "Is there not some way of handling this matter which will do justice to everyone concerned and at the same time not appear in the literature of physics in the United States as a personal controversy between men of such eminence as you and Compton," Tate implored.[15]

Millikan apparently received Tate's letter just after he had dispatched his response to Compton and was also very much perturbed. "It looks as though I should have to withdraw the one element of commendation contained in my letter to [Compton], since he apparently sent it on for publication without waiting to get together with me upon the accuracy of his attack," he wrote to Tate. (Compton had, in fact, waited about ten days for a response from Millikan before sending the articles to *Physical Review* and requesting "prompt publication."[16]) Nevertheless, Millikan suggested that the "simple, dignified, and correct way" to handle the situation would be to "get some disinterested person" to write a review article on the history of the subject at hand. Perhaps the person could even visit the labs of both Millikan and Compton, with expenses equally paid by both, to gather information. "It would take much less time than defending ourselves from incorrect attacks," wrote Millikan. "Let me know whether you think any such scheme is feasible before I have spent a lot of time pointing out errors to Compton himself."[17]

Compton continued to maintain a cool professionalism in his communications to all concerned. "May I thank you for your courtesy in expressing your reaction to the proposed articles by Clay and myself," he wrote to Millikan, adding in a touch of polite understatement, "It is evident that my paper gave to you an impression of being much more personal in character than I had intended that it should." He was in "complete agreement" with the suggestion that some disinterested third party write an account of the whole matter for the *Reviews of Modern Physics*, suggesting that Bell Laboratories physicist Karl Darrow might be a good candidate, if Millikan agreed, or perhaps someone else suggested by Tate. Meanwhile, Compton agreed to withhold publication of the papers until the situation could be amicably resolved.[18]

Some back and forth then ensued on suitable candidates to take on the job of disinterested referee/cosmic ray historian, mostly centering on either Darrow or the Bartol Research Foundation's Thomas Johnson, whom Tate noted was already working on a summary of cosmic ray measurements for the *Reviews of Modern Physics*. But the matter was not to be settled quite so quickly. As the new year of 1937 began, Compton informed Tate that he had finally heard back from Clay in Amsterdam, who was quite enthusiastic about Compton's paper and his addition: "As far as I know, all the physicists working in the cosmic ray field will approve of our action."[19] Because of this, Compton said, "it is apparent that Clay will be disappointed if these papers are not published."

As an alternative, if Darrow or Johnson were unable or unwilling to write a neutral review, Compton stated "I see no reason why the main difficulty which Millikan mentions in connection with these papers—that they are an 'argumentum ad hominum'—cannot be avoided." Compton could revise and shorten his original paper "in such a way as I believe to avoid this objection." Perhaps, then, in order to save time and further trouble, Compton could send the revised papers to Millikan with the invitation to respond as he saw fit, and all would be published in the same issue of *Physical Review*. Tate, who had not yet had a chance to approach Darrow on the matter, agreed. Compton sent a copy of his letter on to Millikan.[20]

Unfortunately, the tetchy Millikan was having none of it. "I think the correct solution is the one to which you have already agreed [i.e., the independent outside party reviewer], and that the solution which you suggested to Tate is entirely incorrect [i.e., a revised Compton/Clay paper with a Millikan reply]," he responded to Compton on January 19. "Incorrect, because you think I have shown my incapacity to deal properly with your work; and I am certain that during the last five years you have demonstrated your incapacity either to understand my work or to handle it in a way which seems to me to possess the first element of correct scientific treatment."

Despite Millikan's repeated outcries about Compton's "incorrect personal attacks" and ad hominem arguments, his letter is itself eight pages of the most direct, angry, vitriolic personal attack, covering "many facts which should now be forgotten as bygones," rebutting the various points regarding the latitude effect and the history of its discovery that Compton had covered in his article.

"You perhaps think you can eliminate personalities and discuss scientifically your results and mine, but the evidence seems to me overwhelming that you cannot do so," Millikan wrote, in a sentiment that could more accurately be expressed of himself. "Practically every page of this manuscript contains the same kind of misrepresentation or other use of the 'smeer [sic] Hoover' type of technique." Millikan concluded his diatribe by observing that "much more might and indeed should be written if this stupid and demeaning controversy must be continued; but I am quite sure that your reputation will be enhanced if it is brought to an end and our original plan of asking Darrow to write up the history of the latitude effect is followed."[21]

Millikan's January 19th letter was "a splendid illustration of the dictum that many small truths can result in a big lie," noted De Maria and A. Russo. "Millikan was certainly right in many details. But he could not wipe out the fact that his research had been guided by an erroneous preconception and that his acknowledgment of others' work stood at or beyond the threshold of scientific correctness. Compton was not wise enough to attack these general aspects, and so left his claims about priority and his criticism of specific data open to challenge from Millikan's punctilious references to previous papers and statements."[22]

Just as Millikan had done with Compton's paper, Compton scribbled some notes to himself on Millikan's new letter. Unlike Millikan's, however, Compton's few handwritten notes were free of piqued remarks such as "simply ridiculous!," merely noting passages and references that he wanted to check on or clarify in any future response. The contrast between Compton's sparse, cool commentary and Millikan's affronted reactions is striking.[23]

After writing to Tate with his new proposal for a revised article, Compton had also written to Clay to fill him in on the latest developments, and because of the mail delay to Europe, received a response at the beginning of February. Clay reiterated that he would "very much regret" if publication of their article was delayed and thought that as long as it was presented objectively as a scientific contribution rather than a personal attack, it didn't matter what Millikan thought of it. "I think Millikan tries to prevent the publication just because it is difficult for him to refute it, for he is militant enough." Clay also objected to the idea of Thomas

Johnson as a disinterested party, believing him to harbor some personal grudge against him, and also had some reservations about Darrow's suitability for the role.[24]

That posed a problem, however, because in the meantime, Millikan told Tate that as far as he was concerned, "Further personal discussions by him or by me of the others' results can serve no useful purpose . . . neither he nor I should be commissioned to attempt to make objective statements which neither of us thinks that the other is capable of making." For that reason, "The only satisfactory way to handle the problem is to go through with the Darrow plan as agreed upon."[25] Millikan's papers also contain fifteen pages of undated handwritten notes filled with anger in which he attempted to work out his various objections and to detail the "attacks" he believed he had suffered. These were certainly only for his private use and not for publication, but they make abundantly clear that for Millikan, the entire affair, not simply the present situation involving *Physical Review*, was a deeply, intimately personal matter.[26]

Instead of a swift resolution, the situation seemed to have run into a wall of intransigence on all sides—echoing the larger Millikan-Compton conflict that by now was already at least five years old. But likely because the conflict had now become both sharply focused on a specific venue and was also free of the messy complications of press involvement and public awareness, it was also building to at least some sort of resolution.

Millikan next approached Compton with what he apparently thought was a reasonable reaction. He remarked that he had "written out replies to what seem to me the most glaring of the misstatements of fact, history, and interpretation contained in the article which you sent on to Tate," apparently referring either to his January 19th letter or to the notes he had worked out. Now, however, since Compton had "played good ball" in withdrawing his article, Millikan decided "it will now be better to bury the whole matter in my files." Because an attempt by Compton to revise his article would be "essentially a return to the method of personal controversy," they should instead proceed with the third-party solution already agreed upon. "This is all that the scientific world is or should be interested in," he wrote. "It is only the vulgar populace that is interested in a dog fight."[27]

Now it was Compton's turn to be stubborn. Tate still had not approached Karl Darrow on the situation and in fact was doubtful whether he would want to put himself in the middle of the argument. Apparently out of patience, Compton decided to force the issue and proceeded to write a revision of his paper that would presumably be less of an "argumentum ad hominum." He sent the revision to Millikan, saying that "I decided that the undertaking of preparing a paper discussing the present situation of the geographical study of cosmic ray intensity really should fall to me rather than to someone else." Still, Compton remained unfailingly cordial: "May I thank you for counseling me regarding the form which this paper should take. I feel that it is a much more worthy and valuable statement as a result of your comments."[28] Tate invited comments from Millikan, observing, "Apparently Compton feels that it is impracticable to find anyone who would be willing to undertake an independent review of this subject."[29]

Millikan seemed to take this as yet another opportunity to play the aggrieved victim, the noble, injured party who had done his best to stay above the fray in an intellectual high tower, while his adversary continued to sully his good name and drag him down into the mud in every way possible. He suggested to Tate that the journal adopt a policy of "asking the participants to a controversy to work out their differences *privately*, rather than publicly," either with the journal editors or an outside party chosen by them. The "attacking article" could simply be sent back and forth between the "contestants" for revision until both were satisfied, or until the "referee" decided it was correct, before finally publishing.

In the present case, said Millikan, "there is no understanding Professor Compton's present attitude and action save on the supposition that he has actually convinced himself that the Pasadena group has been not only very incompetent but wilfully [sic] dishonest in its presentation. . . . He has spread this view rather widely, even to soliciting and securing Clay's support in his attack." Because Millikan had thought that the Compton-Clay piece had been withdrawn, he kept his reply to himself and "buried it in my files." But Compton changed his mind and resubmitted "a very considerably revised and improved article, but still a very one-sided one, definitely incorrect in a great number of particulars." Millikan enclosed a seven-page response to Compton's article. He stuck to the plan of enlisting some impartial referee who would "get Dr. Compton to revise his article until it is historically and scientifically correct."

As far as Millikan was concerned, "for the dignity of our American physics this demeaning, and I think entirely one-sided, controversy in which you know I have not participated at all, should come to an end at least as far as the public is concerned." In a gesture of apparent generosity, in case the referee plan wasn't or couldn't be pursued, then Millikan "shall be quite content if you publish Dr. Compton's article as it is," meaning the new revised version. However, Millikan couldn't resist adding a comment to Tate that "It would have been better for me personally had you published the first one without suggesting that he change it at all," either forgetting or ignoring the fact that it had been Compton who had offered to both revise the paper and withhold its publication. But Millikan was "glad for his sake that our discussion has improved it somewhat. If the facts I am sending to you and to him today lead either of you to take steps toward further improvement, well and good." If not, however, then the passage of time and further work by others "must be relied upon for rectifying the incorrectness."[30]

The "Comments on Arthur Compton's Paper" that Millikan sent to Tate and Compton were far more restrained and professional than his more personal responses, but no less defensive and argumentative. Compton made various notations on his copy, again neutral and not at all vindictive in tone, even noting "my mistake!" and "thanks for information!" in some passages. He sent Millikan a brief note thanking him for his comments on the article revision. "I am glad that you have mentioned the specific points with which you take issue with my statements, for I feel that this gives us a basis on which our differences may, to at least a large extent, be ironed out. It may be some time before I am able to send you a detailed reply, but I hope that when this is done you will see that we are beginning to find a basis for a better appreciation of each other's position."[31] He again asked Tate to withhold publication for the time being.

With that, the matter lay fallow for several months. Not only were all the participants understandably weary from the struggle, but there was other work to be done, trips to be made, lectures to be given, cosmic ray instruments to be flown on airplanes and balloons and installed on ships and mountaintops.

At last, perhaps reaching a final and profound understanding of the limitless depths of Millikan's stubbornness and the futility of trying to erode its implacable granite facade, Compton completely withdrew his *Physical Review* paper, revised or not, and suggested to Clay to do the same. In August, he informed Millikan of the decision.

"When I realized that the publication of my article might really hinder the effectiveness of your work, I could not but agree that its appearance was inadvisable," Compton wrote. "The point you make regarding avoiding publicizing 'quarrels in the family' is a cogent one." Once more, he thanked Millikan for providing "a better understanding of the ideas underlying some of your early experiments."

Compton made clear that "I still retain the feeling that your presentations would have been greatly strengthened if they had included something like an adequate statement regarding the latitude effect studies of others," but he allowed that "it is obviously justifiable to confine oneself to a presentation of one's own findings." If he harbored any ill will toward his senior colleague, it was not evident in Compton's laudatory closing: "When, as you have done, a man can go ahead with first class experiments and clear thinking, undismayed by criticisms of others and without undue prejudice by his own earlier attitudes and statements, and thus greatly enhance the understanding of nature, he deserves the high regard of his colleagues. I hope that no minor differences which may arise between us will cause you to doubt the sincere esteem and admiration which I thus hold for you."[32]

Perhaps Compton was laying it on a bit thick, but Millikan nevertheless seemed to be more or less satisfied. "My dear Arthur," he responded, "I am of course glad that you made the decision which you did, and I think it was a wise one."[33]

As De Maria and Russo noted, Compton had come to the conclusion that he had no good options to proceed any further. "At this point Compton had no choice but to withdraw or expose the American scientific community to a public controversy that could only damage its prestige."[34] By this time, both he and Millikan had already experienced firsthand far too much of the nastiness and intensity of a scientific disagreement played out in the public sphere and plastered relentlessly over the national press. Aside from damaging their own personal prestige and that of science in general, such a furor wasted time and energy that could be better spent doing science, instead of providing reporters with juicy quotations.

Compton likely decided that his best course now was simply to concentrate on and continue his own work and let the results speak for themselves and allow the weight of scientific consensus to continue to tilt in his direction, as indeed had already been taking place since 1932. "Ultimately Compton had neither the endurance nor the appetite for what he began to see as a counterproductive squabble," noted Millikan biographer Robert Kargon.[35]

Another factor was undoubtedly simple fatigue, especially for Millikan, who was about to turn seventy years old and preferred to spend his remaining years doing science, not providing entertainment for the "vulgar populace." He was, as always, also alert for any threats to his dignity and reputation, and the years of controversy with Compton had left him even more keenly sensitive to any possible revival of the unfortunate affair.

Toward the end of 1937, he wrote to Arthur H. Sulzberger, publisher of the *New York Times*, to take exception to an article. Complimenting the usually "accurate and dignified" science reporting as exemplified by Waldemar Kaempffert and William Laurence, he explained that he was "quite loath to call your attention to a very unfortunate and unworthy case of scientific reporting" regarding a lecture he had recently delivered at Columbia University. He felt the reporter had misrepresented various statements, particularly the characterization that his new findings "took vigorous issue . . . with an opposing school of thought, headed by Professor A. H. Compton." The *New York Times* reporters, Millikan suggested, might stop trying to "create the impression of a dog fight going on between physicists." He hadn't even mentioned Compton in his lecture, he complained. In a postscript, Millikan praised Laurence's coverage on a different subject, emphasizing that "it covers much the same ground but does not play up *non-existent controversy*."[36]

Sulzberger responded that "Mr. Laurence was out of town, so that another man who was less familiar with the subject had to be assigned . . . his introduction of the Compton controversy was by way of supplying background to clarify the story for the average reader, but there was no intention of making it appear that you had mentioned him in your address. It really boils down to a case of unskillful coverage, which we deeply regret."[37]

As far as Millikan was concerned, the conflict was over and done with, and he was not going to allow it to be revived, at least publicly—even if he continued to disagree with Compton's ideas. But if Millikan and Compton had achieved a sort of peace between themselves, however, Jacob Clay remained not at all pleased. He wrote a long letter to Compton complaining about Millikan and his cosmic ray claims: "I am forced to have the worst idea of the character of Mr. Millikan as I have seen that he is violating the truth, as he does, for his own profit, without any scruples."[38] Compton informed Tate that although Clay agreed that it was too late to publish the papers he and Compton had prepared, Clay was "more emphatic than ever in his feeling that Millikan's papers have done him considerable injustice."[39]

Much to Tate's dismay, he would be forced to confront another potential Millikan controversy a couple of years later in 1940, when Clay submitted to him a critique of a new collection of Millikan cosmic ray lectures. Clay's review was decidedly far more of a direct ad hominem attack than anything Compton had submitted earlier. "From the beginning of his work on cosmic rays up to 1933 [Millikan] has unfortunately been on the wrong side with nearly all his own experiments and conclusions," wrote Clay. Millikan "tried to turn the facts in such a way that they prove that he has been right, and partly he tries to attain his end by distorting the results of others and depreciating their work."[40] No doubt wanting to avoid another prolonged conflict and again stirring up Millikan's ire, Tate declined to publish Clay's piece.

The Chief, it seemed, had won again. But only apparently. As Compton's conciliatory letter had observed, "The historical matter will in time take care of itself." Millikan and Compton had achieved at least an armistice, if not a full peace treaty. But the final referee in their conflict would prove to be history itself.

18

A SORT OF PEACE

By the late 1930s, the glorified parlor trick of using the detection of cosmic rays to activate loudspeakers or light displays and wow an audience had become something of a fixture, particularly in public lectures on the subject. Even Millikan and Compton made use of it on occasion, finding it a useful way to demonstrate the reality of cosmic rays to a crowd of the general public. But the most spectacular example was probably the display opening the New York World's Fair on the evening of April 30, 1939. Or at least, it would have been spectacular, if it had worked as planned.

The opening ceremonies were supposed to be "hullabaloo and ballyhoo seldom surpassed in this hemisphere," according to the *Chicago Tribune*.[1] Atop the 610-foot-high Trylon, the needly spire next to the 180-foot-diameter Perisphere that together symbolized the Fair, a series of ten flashes of light were to be activated by the detection of ten cosmic rays by a device centered on a triple-coincidence Geiger-Müller counter built by the Bartol Research Foundation's W. F. G. Swann. Albert Einstein, now a U.S. resident and soon to become a citizen, had agreed to give a brief address in person and on the radio explaining cosmic rays, though reluctantly. "Dr. Einstein protested when the project was first broached to him that it would take a volume to begin to elucidate the subject," reported the *New York Times*. He managed to write a speech of about seven hundred words, apologizing to World's Fair officials that he simply couldn't cut it down any further.[2]

"If science, like art, is to perform its mission truly and fully, its achievements must enter not only superficially, but with their inner meaning, into the consciousness of people," Einstein intoned in his thick German accent, battling a faulty sound system and bad acoustics that made his speech almost indecipherable to the crowd. "To serve this purpose, I will use this occasion of the opening of the New York World's Fair to describe very briefly how the physicists came to discover one of the most remarkable of all natural phenomena, and one of the most mysterious in its origin, known by the name of 'cosmic rays.'" Einstein then proceeded to give a quite brief synopsis of the topic, including the early balloon flights. Whether because he had to keep things short, or simply to avoid possible controversy, the only name he mentioned was Victor Hess, not Millikan or Compton.

After Einstein's speech ended, an announcer declared, "We will now call on these interplanetary messengers to reveal the World of Tomorrow," explaining that the rays were just then traveling toward New York at 186,000 miles per second. "Give us ten cosmic rays," he implored the universe. "We will make the rays visible and audible to you, and on the tenth ray, the World of Tomorrow will be illuminated." As each ray was captured, a ring of bright lights flashed around the pinnacle of the Trylon, each accompanied by a bell tone from a six-foot piano wire, designed to sound like a bell thirty feet in diameter.

The tenth ray was supposed to set off the brightest flash of all, equivalent to a million 100-watt light bulbs and visible for twenty miles, lighting up the entire fairgrounds. But the power of cosmic rays, or more likely the strain of the sudden electrical surge, proved too much. "Something happened to the electrical system when the switch was thrown to illuminate the last huge light on the Trylon," said the *New York Times*. "Instead of making the Fair as bright as day with the 'Star of the Trylon,' it put some of the lights out." A deflated crowd moved on to a magnificent display of colored lights and jets of water in the Lagoon of Nations. "The crowd dropped science in favor of a spectacle they could applaud."[3]

Cosmic rays were still a good way to draw a crowd and to fill column space in newspapers, but the emphasis now was mostly on their inherent awe and mystery, with articles regularly trumpeting news such as "Cosmic Ray Puzzle Found," "Cosmic Ray's Source Traced," "Cosmic Rays' Secret Key Believed Found," "New Cosmic Ray Secret," and speculations about their use for "limitless power" and similar fancies. The subject continued to be popular in various advertisements and among medical quacks and self-styled religious figures alike. Even the famous H. G. Wells couldn't resist the lure of the cosmic ray, centering his 1937 novel *Star Begotten: A Biological Fantasy* on the idea of an invasion of Earth undertaken by Martians altering the minds and chromosomes of humans by remote control through the manipulation of cosmic rays.

But cosmic rays no longer provided the focus for a titanic battle of Nobel giants. Robert Millikan and Arthur Compton remained in the public eye, their various travels and speeches and new experiments duly noted, especially in their respective hometown press. They would frequently be seen at the same conferences and confabs, such as an international cosmic ray symposium at the University of Chicago organized by Compton in June 1939. Each continued his own research programs, Compton setting up instruments on Colorado mountaintops and elsewhere and up in his automated balloons, Millikan embarking on a voyage to Australia and India in the summer of 1939 to conduct still more measurements, despite the fact that at age seventy-one, he was getting somewhat long in the tooth for globetrotting.

The advent of World War II in September 1939 didn't affect their work too severely at first. But as the war spread, intensified, and finally engulfed the United States in 1941, most scientific work not directly related to military purposes was effectively suspended, whether because of the redirection of funding, lack of ready manpower, or lack of necessary materials. With Europe and its scientists and resources cut off for the duration, the casual give and take of international scientific collaboration ceased. Although some work in cosmic rays and physics still went on even in parts of Europe despite the war, researchers couldn't communicate it to each other.

Unlike his involvement in the National Research Council in World War I, Millikan's advancing age, along with his strongly anti-Roosevelt and anti-New Deal predilections, kept him out of any direct involvement this time, his contributions to the war effort limited mainly to keeping

Caltech and its facilities, particularly its aeronautical research department, fully running and supplied with resources and necessary personnel. Arthur Compton, however, became intimately involved in the war as a major figure in the Manhattan Project that would build the world's first atomic bomb. As director of the Metallurgical Laboratory at the University of Chicago, he presided over the creation of the first nuclear reactor, an essential step to produce the bomb's raw material, and later became one of the chief members on the presidential committee advising on the bomb's use. For Compton, whose interest in cosmic rays was at least initially borne of his fascination with the mysteries of the atomic nucleus and the structure of the atom, he was now perhaps more directly involved in probing and considering those mysteries than he might have imagined or wished during his years of lugging electroscopes up mountainsides in South America or flying them over the frozen Arctic.

Robert Millikan finally officially retired as the de facto president of Caltech in 1945, but characteristically refused to retreat quietly into dotage. He continued to do research, travel, and make public pronouncements on matters of science, politics, and philosophy.

But his time had long passed. As Robert Kargon noted, "On scientific matters, Millikan was revered but not heeded." In 1950 he published a lengthy autobiography. "The self-portrait that Millikan painted parallels closely the familiar picture of the American pioneer . . . self-reliant, hardworking, and determined," Kargon wrote. Millikan's proud conception of himself as an independent, self-made man was on full display. "A grand old man had made his statement—a little rambling, a little unbalanced, but still a piece of Americana. There were no critical reviews. He had become, after all, a monument."[4] Millikan died in December 1953 at age eighty-five, just a few weeks after Greta, his beloved wife of over fifty years.

Arthur Compton, feted for his major contributions to the Manhattan Project that had ended the war, finally left Chicago in 1946 to become Chancellor of Washington University in St. Louis, the same institution where, decades earlier, he had conducted his Nobel Prize-winning experiments. Like Millikan, he too would continue his research and steadily move into a role as an elder scientific statesman and sage. He would become a major force in establishing Washington University as an important research institution. After retiring in 1961, he died in March 1962 at age sixty-nine. He published many books and popular articles, though unlike Millikan, the closest he ever produced to a full autobiography was a 1956 account of his Manhattan Project experiences entitled *Atomic Quest: A Personal Narrative.*

In late April of 1920 at the annual meeting of the National Academy of Sciences at the Smithsonian Institution in Washington, D.C., long before Robert Millikan and Arthur Compton faced off before a crowd of colleagues and reporters in Atlantic City, another titanic debate occurred about some big scientific questions and the nature of the universe itself. Known among astronomers as the "Great Debate," the participants were Harlow Shapley from Mt. Wilson Observatory in Pasadena and Heber Curtis from Lick Observatory in Northern California, and the main topic was "The Distance Scale of the Universe," basically the question of whether the so-called "spiral nebulae" were within our own galaxy or were far more distant "island universes" of their own, meaning, separate galaxies. Shapley argued the former position; Curtis the latter.

Despite the reputation it has assumed with the passage of time, however, the "Great Debate" was hardly noticed or remarked upon at the time outside of the academy. As science historian Marcia Bartusiak explains, "In astronomy circles, the venerable legend that surrounds that April session—the memory of it as the mighty clash of cosmic titans, astronomy's version of *High Noon*—developed gradually over time, the embroidery added so profusely over the years that it was eventually described as a 'homeric fight,' two opposing sides battling it out in the highest court of scientific opinion." Hardly a debate in the true sense of the word, the NAS session was really "more like two lectures back to back."[5] A year later, however, Shapley and Curtis would have more of an actual debate in the scientific literature, trading papers and arguments, and Bartusiak points out that it was actually this written back-and-forth that became conflated with the NAS presentation, establishing the abiding legend of the "Great Debate."

It's unknown whether Millikan or Compton personally witnessed the "Great Debate," although Millikan, at least, was present at the three-day NAS meeting to deliver a paper. But although the Shapley-Curtis debate was probably far more significant scientifically with the distance of a century's hindsight, the Millikan-Compton controversy far surpassed the Great Debate in its contemporary prominence and definitely its bitterness. In the Great Debate, the nature and size of the universe were specifically up for grabs; but while Arthur Compton was concerned mainly with the specific outstanding questions on cosmic rays, Robert Millikan saw the stakes of his contest with Compton to be nothing less than the fate of the universe.

Both men were deeply spiritual, both were religious in their own ways, but Millikan was more invested personally in the controversy, and that along with his ingrained stubbornness and ego were the major reasons that the Millikan-Compton conflict dragged on for so long and became so heated at times. The Shapley-Curtis debate never became so personal and even at its most intense, retained the more or less friendly, or at least professionally cordial, character that the press sometimes liked to claim of the Millikan-Compton affair. The questions at issue between Shapley and Curtis also never became so personally identified with each man but rather, remained scientific hypotheses that each happened to be advocating. And when those questions were eventually resolved with further scientific evidence, partially vindicating the positions of both men, the matter was set aside and largely forgotten by both Shapley and Curtis and rarely if ever again mentioned.

Harlow Shapley and Heber Curtis, however, never had to deal with the same tsunami of press coverage and commentary that accompanied every twist and turn, every new development and finding, every meeting and encounter, of the Millikan-Compton saga. Because of the celebrated status of both its participants, especially Millikan, the discussion from its earliest stages was inevitably fated to become public, but as it proceeded and expanded and intensified, it became harder for both Millikan and Compton to ignore, especially when it reached the point where they were each routinely confronted by reporters with direct questions about it and about each other. Eventually it affected both men and their approaches to their work and to each other. The somewhat more thin-skinned and sensitive Millikan seemed in particular most affected and influenced by the public aspects of the affair, as evidenced by his increasingly frequent comments on the inaccuracy or sensationalism of newspaper coverage. It also didn't help that noisy public spectacles such as stratospheric balloon flights and World's Fairs served as a backdrop for the very questions that Millikan and Compton were battling to settle.

Such considerations may have actually contributed to the tacit, somewhat incomplete resolution of the conflict that occurred in 1936–1937 with the *Physical Review* episode, which Millikan was quite anxious to keep firmly out of the glaring, unforgiving, judgmental public eye, appealing to Compton to help keep private any "quarrels in the family." It was not simply a matter of avoiding the annoyances of blaring newspaper front pages and smirking editorial columns, but also preserving the dignity and effectiveness of physics and the process of scientific research in their essential form. That was something that both Millikan and Compton realized was not only in their own best personal and professional interests but also those of their colleagues and their entire profession, and which may have finally been the decisive element that convinced both men to withdraw from further battle, set aside their differences, and leave whatever outstanding questions remained to the verdict of history and the scientific process.

In many ways, unlike the Great Debate, the Millikan-Compton feud was in the end a war of attrition, with no firm victory or conclusion. By 1936, the weight of evidence that primary cosmic rays were indeed charged particles, mostly protons and helium nuclei, with a small percentage of photons and an exotic array of new particles produced in the form of secondary ray showers, had become so overwhelming that even a giant like Millikan simply couldn't deny it any further. But he didn't have to like it. Even as the Millikan-Compton feud sputtered out from weariness on the part of the participants and ever-increasing scientific consensus favoring Compton's position, Millikan never really ceased his efforts to find some way around it, to prove somehow that at least in some way, in some part, he had been right after all.

Settled or not, echoes of the debate persisted for years afterward, and not merely in scientific circles. Writing to Compton in 1943 from a Carnegie Institution cosmic ray station in an Echo Lake, Colorado, army camp, researcher Marcel Schein described a query from one of the army men: "He spoke very highly of you and his first question was whether I am on your side or Millikan's side in cosmic rays. I tried to explain to him that we have now in Chicago an entirely new hypothesis regarding cosmic radiation which is quite different from Millikan's old ideas. It is really astonishing that one finds here such an interest in cosmic rays."[6]

And yet, almost paradoxically considering how intense the debate sometimes became, Millikan and Compton also each maintained a true scientist's attitude with a willingness, however reluctant at times, to change his mind and alter his hypothesis when the evidence demanded it. Millikan first displayed that quality in 1926 when, confronted with the new data he had been gathering for several years, he accepted the extraterrestrial origin of the "mysterious radiation" after first vehemently arguing against it, reversing his position to the extent that he began calling it "cosmic" rays (whether or not others had already first used that term). Much later, he finally accepted, at least publicly and for all practical purposes, the particulate nature of the rays and the latitude effect, no matter how much he may have also struggled to force and tailor his beloved birth cries theory to fit each new discovery and bit of data.

Compton also showed the same scientific openness, certainly more readily than Millikan, in altering his own views when evidence made it necessary: his ideas of extragalactic vs. galactic cosmic ray origins, that cosmic rays largely originated within the atmosphere rather than from outside the earth, that photons did indeed form some component of the entire cosmic ray flux. When he first shifted the emphasis of his research to cosmic rays, he also displayed a considerable degree of professional and personal courage in bucking what was then the established wisdom and setting himself up directly against the titanic figure who exemplified and embodied that accepted wisdom, Robert Millikan. The fact that he also had a Nobel Prize may have provided Compton with something to steel his resolve, but in the early 1930s, it still took a considerable degree of nerve for a younger and not-quite-as-well-established scientist to directly confront the most senior, most famous, and most prestigious physicist in America on a fundamental question of science in which he was universally recognized as the ultimate authority.

The Millikan-Compton debate was perhaps the first major public example of just how heated the controversies among scientists could become—and the lengths that scientists would be driven to in order to settle their disputes. As such, it's not only important for the scientific questions involved, but also because, then and now, it serves as a reminder that however brilliant and exceptional, scientists are still human beings. They have their own personalities, passions, blind spots, and gifts; they can argue, make mistakes, harbor prejudices and misconceptions, and behave like moody children. What sets them apart, at least in most cases, is their ability to eventually set aside those baser and less enviable tendencies and finally concentrate on what matters in the end: the science. Robert Millikan and Arthur Compton never became best buddies after their famous clash, but each man did his best to find the truth, whether or not it supported his own beliefs.

Beyond the work, personalities, and idiosyncrasies of Robert Millikan and Arthur Compton and the specific subject of cosmic rays, the entire episode provides some other significant insights as a case study of how science works. A unique phenomenon, in this case a mysterious radiation from an unknown source, is discovered; questions are asked, hypotheses formed, and experiments devised and carried out to test them and to find answers. Some answers are found while new questions are raised, and the process continues, each step bringing further hypotheses and questions, until theoretical frameworks begin to come together to explain it all.

Because science is also a human process carried out by fallible human beings with egos and personalities and quirks and their own individual beliefs, hypotheses and theories and experimental directions are sometimes unduly influenced, and effort is wasted in unfruitful directions. Sometimes researchers become too enamored of particular ideas or explanations in ways that blind them willingly or unwillingly to alternatives: for example, Millikan's devotion to the idea of elemental birth cries and continuous creation in an eternal and unchanging universe.

The necessary multiple efforts by different people using different methods, many of them jockeying for credit and recognition and resources, can slow and complicate the process, but as long as the process is allowed to continue unfettered by outside ideological, political, or philosophical restraints or by attempts to make science fit a preferred view, a unified effort develops, and consensus is established. Yet the consensus is never completely fixed and set in concrete; it must always be allowed to evolve and change as new facts and data are gathered and a true and valid picture of reality is achieved, adding new knowledge to the collective consciousness of humankind.

From the first inklings of their existence through all the research, work, arguments, and theories that have been accomplished and advanced until the present day, cosmic rays have provided a defining example of the unique nature of science as a means and a tool for understanding the universe and our place in it. Even with the unavoidable personal conflicts and complications that arise in every realm of human endeavor, the eternal expanse and indifference of the universe supersedes our own pettiness and reminds us both of our ultimate smallness and our tiny, yet nevertheless real part in all of it.

In the words of film director Stanley Kubrick, creator of *2001: A Space Odyssey*: "The most terrifying fact about the universe is not that it is hostile but that it is indifferent, but if we can come to terms with this indifference, then our existence as a species can have genuine meaning. However vast the darkness, we must supply our own light."[7]

POSTLUDE: THE UNITY OF ALL CREATION

In his 1964 book *Cosmic Rays*, Bruno Rossi, the pioneering cosmic ray researcher who among other contributions had developed the coincidence circuit so important in the field and also discovered the phenomenon of extensive cosmic ray showers, made a rather reluctant, somewhat rueful admission. "Half a century after the discovery of cosmic rays the problem of their origin is still unsolved," he wrote. "We do not know for certain where cosmic rays come from. We do not know for certain how they acquire their tremendous energies."[1]

More than half a century after Rossi wrote those words, they remain just as true. Although we've learned much about cosmic rays since their discovery more than a century ago, we still don't know exactly where they come from. Cosmic rays remain one of the most intractable scientific puzzles of all time.

The post–World War II era brought a blossoming of new findings, thanks largely to wartime innovations such as high-altitude rockets, more sophisticated detectors, and eventually Earth-orbiting satellites and deep space probes that allowed data collection outside of the atmosphere for the first time.

A new generation of physicists eagerly took up the cosmic ray banner, with researchers such as James Van Allen and John Simpson using increasingly sophisticated rockets and balloons to probe the edge of outer space and beyond in ways that Millikan and Compton could only dream of.

Such work soon revealed other varieties of cosmic rays, strange new particles, and new clues as to their composition. Their explorations were complemented by the work of veteran researchers such as Enrico Fermi, who came up with an important new theory on a possible origin for the rays, and the advent of powerful particle accelerators that allowed cosmic ray phenomena to be recreated and tested. NASA and a dedicated space program began to send satellites and probes beyond Earth, providing more data and inspiring new questions by allowing cosmic rays to be detected and studied in space for the first time. Biennial international cosmic ray conferences began a tradition that continues to the present day.

In the 1960s and the following decades, the discovery of the microwave background radiation revolutionized astrophysics and theories about the origin of the universe, with profound implications for cosmic ray physics. New observations with increasingly sensitive and versatile detectors and spacecraft are unveiling a cosmos permeated by cosmic rays of unimaginable energy, hinting at the inner workings of supernovae and black holes. The detection in 1991 of the most powerful cosmic ray event ever recorded, known as the "Oh-My-God particle," sparked an ongoing search for similar events, and raised curiosity about their possible causes. International efforts were organized to construct huge arrays of cosmic ray detectors, such as the Pierre Auger Observatory in Argentina. Cosmic ray science expanded in a synergistic relationship with astronomy, particle physics, radio astronomy, and new fields including neutrino and gamma-ray studies.

In the twenty-first century, now well established as a vital discipline in astrophysical research, cosmic ray science has helped to spawn a new revolution called multimessenger astronomy. By combining observation and data collection from the entire range of the electromagnetic spectrum in ways never before possible, multimessenger astronomy is providing fresh lenses through which to study and understand the universe. Traditionally separate fields of physics and astronomy are merging to provide a way of seeing the universe as a whole, rather than focusing on a tiny sliver of it as in the past. Instead of seeing the universe only through visible light, or radio waves, or infrared radiation, or another limited piece of the electromagnetic spectrum while ignoring all the rest, multimessenger astronomy allows astronomical objects and phenomena to be observed simultaneously through a multiplicity of windows, so that the intricately woven relationships of the processes and energies that create and drive them can be perceived *in toto*.

Cosmic rays and the phenomena associated with them are a brightly colored thread running through and tying together a broad range of formerly disparate disciplines, strengthening their insights while providing a framework for the most profound questions facing twenty-first-century astronomy, including the existence and nature of dark energy and dark matter; the formation of galaxies; the fundamental nature of matter; and the ultimate fate of the universe. Meanwhile, a final answer to the question that has perplexed scientists for over a century still remains elusive: where *do* cosmic rays come from?

Cosmic rays may not provide humankind with the "secret of Creation," but they continue to bring us closer to it, or something like it. They may not be "the birth cries of atoms being born in interstellar space" or their "death wails," but they can provide deep insight into the heart of black holes and galaxies at unimaginable distances, the existence of dark matter and dark energy, and the birth of the universe. Perhaps, in ways they never could have imagined, Robert Millikan and Arthur Compton were right after all. We, like cosmic rays, are ourselves splinters of infinity.

ACKNOWLEDGMENTS

I'd already known that librarians were special beings with wondrous powers since at least the second grade, when they led me to all the cool science and science fiction books in the school library. But I was given a whole new reason to treasure librarians during the global COVID-19 pandemic, which stopped travel and closed down the world, preventing access to libraries and archives. Like so many other authors, I was forced into a new and severely limited way of working during the pandemic, but thanks to the above-and-beyond efforts of several dedicated librarians and archivists, I was able to obtain remotely and electronically the material I needed to complete this book. Much gratitude to Alison Carrick and Sonya Rooney at Washington University in St. Louis; Joe DiLullo at the American Philosophical Society in Philadelphia; Penny Neder-Muro at the California Institute of Technology in Pasadena; and the staff of the MIT Libraries in Cambridge, Massachusetts.

For interesting and helpful discussions on cosmic rays, Millikan and Compton, and the universe in general, thanks to David Kaiser of MIT; Wystan Benbow of the Harvard-Smithsonian Center for Astrophysics and director of the VERITAS Observatory; David B. Kieda at the University of Utah; Marcia Bartusiak; and Deborah Blum.

For moral support, friendship, and help both tangible and intangible, thanks to Jeff Harris, Suzanne Rosin, Kristina Finan, Linda Chamberlain,

Sue Smith, Cat Calhoun, Judy Weightman, Denise Shubin, and Maura R. O'Connor. Thanks also to the marvelously talented Genevieve Leonard for her wonderful artwork.

My research and writing of this book were supported by a grant from the Alfred P. Sloan Foundation. Thanks to the foundation and to program director Doron Weber for their generous support of this project.

Finally, many thanks to my agent, Michelle Tessler, my editor Jermey Matthews, Haley Biermann and Kathleen Caruso at the MIT Press, and copyeditor Julia Collins for helping to make this book a reality.

NOTES

PRELUDE

1. Quoted in Abraham Pais, *Inward Bound: Of Matter and Forces in the Physical World* (Oxford: Oxford University Press, 1986), 317; emphasis in original.

2. Fermi would later build on Pauli's work and give the elusive particle a name: the neutrino. It evaded experimental detection until 1956; Pauli died not long after but lived long enough to see his ideas vindicated at last, noting, "Everything comes to him who knows how to wait."

3. Hans Geleijnse, "Mussolini, Intellectuals, and International Conferences: Dutch Conference Participants in Fascist Italy," 2019, https://www.academia.edu/40089082/Mussolini_Intellectuals_and_International_Conferences_Dutch_Conference_Participants_in_Fascist_Italy, 7.

4. Geleijnse, 16.

5. Pais, *Inward Bound*, 317.

CHAPTER 1

1. "Millikan vs. Jeans," *Los Angeles Times*, October 3, 1931, 18.

2. Bruno Rossi, "Arcetri 1928–1932," in *Early History of Cosmic Ray Studies: Personal Reminiscences with Old Photographs*, ed. Yataro Sekido and Harry Elliot (Dordrecht, Holland: D. Reidel Publishing Company, 1985), 53–54.

3. Bruno Rossi, *Cosmic Rays* (New York: McGraw-Hill, 1964), 43.

4. Rossi, "Arcetri 1928–1932," 55–56.

5. Rossi, 56.

6. "The Volta Conference in Rome," *Nature* 128, no. 3238 (November 12, 1931): 861.

7. Rossi, "Arcetri 1928–1932," 69.

8. Quoted in Luisa Bonolis, "International Scientific Cooperation during the 1930s. Bruno Rossi and the Development of the Status of Cosmic Rays into a Branch of Physics," arXiv:1304.5612, Cornell University, https://arxiv.org/abs/1304.5612, 15.

9. Rossi, "Arcetri 1928–1932," 69.

10. Millikan presentation: "Cosmic Rays Called Source of Vast Power," *Chicago Tribune*, October 15, 1931, 19; "Rays Upset Old Ideas," *Los Angeles Times*, October 15, 1931, 1; "Millikan Sees New Notion of Universe," *New York Times*, October 15, 1931.

11. "Cosmic Rays Called Source of Vast Power," 19; "Rays Upset Old Ideas," 1; "Millikan Sees New Notion of Universe."

12. As if she herself were also not one of the "world's greatest physicists"!

13. Curie presentation: "Mme. Curie Explains Rays," *New York Times*, October 16, 1931; "Mme. Curie Explains Cosmic Ray Disclosures," *Boston Globe*, October 15, 1931, 8.

14. "Mme. Curie Explains Rays"; "Mme. Curie Explains Cosmic Ray Disclosures," 8.

15. "Mme. Curie Explains Rays"; "Mme. Curie Explains Cosmic Ray Disclosures," 8.

16. *Nature*, "The Volta Conference in Rome," 861.

17. Arthur Holly Compton, "Personal Reminiscences," *The Cosmos of Arthur Holly Compton*, ed. Marjorie Johnston (New York: Alfred A. Knopf, 1967), 40.

18. Compton, "Personal Reminiscences," 43.

19. Bonolis, "International Scientific Cooperation during the 1930s," 36.

20. Rossi, "Arcetri 1928–1932," 69.

21. Rossi, 69–70.

CHAPTER 2

1. Pierre Auger, *What Are Cosmic Rays?* (Chicago: University of Chicago Press, 1945), 15.

2. V. F. Hess, "Concerning Observations of Penetrating Radiation on Seven Free Balloon Flights," *Physikalishe Zeitschrift* 13 (1912), in *Archives of the Universe: 100 Discoveries That Transformed Our Understanding of the Cosmos*, ed. Marcia Bartusiak (New York: Vintage Books, 2004), 281.

3. Hess, "Concerning Observations of Penetrating Radiation, 281.

4. Hess, 284.

5. Quoted in Mario Bertolotti, *Celestial Messengers: The Story of a Scientific Adventure* (Berlin: Springer-Verlag, 2013), 29.

6. Per Carlson, "A Century of Cosmic Rays," *Physics Today*, February 2012, 31.

7. Michael Walter, "From the Discovery of Radioactivity to the First Accelerator Experiments," in *From Ultra Rays to Astroparticles: A Historical Introduction to Astroparticle Physics*, ed. Brigitte Falkenburg and Wolfgang Rhode (Dordrecht: Springer, 2012), 27. At this point, however, nobody was calling them "cosmic rays."

8. Walter, "From the Discovery of Radioactivity," 27–28.

9. George R. Steber, *Invisible Visitors from Outer Space: Cosmic Rays, The Story from the Beginning to the Present* (Self-published, 2021), 17.

10. Yataro Sekido, "Introduction," in Sekido and Elliot, eds., *Early History of Cosmic Ray Studies*, xiii.

11. Robert H. Kargon, *The Rise of Robert Millikan: Portrait of a Life in American Science* (Ithaca, NY: Cornell University Press, 1982), 74.

CHAPTER 3

1. Kargon, *The Rise of Robert Millikan*, 11.

2. Kargon, 66.

3. Kargon, 66.

4. Kargon, 170.

5. Quoted in Bonolis, "International Scientific Cooperation during the 1930s," 2–3.

6. Robert Andrews Millikan, *The Autobiography of Robert A. Millikan* (New York: Prentice-Hall, 1950), 210. Oddly enough, considering the extent of Millikan's work in the cosmic ray field, he barely discusses it in almost three hundred pages of autobiography.

7. Bertolotti, *Celestial Messengers*, 47.

8. Bertolotti, 53.

9. Kargon, *The Rise of Robert Millikan*, 128.

10. Rossi, *Cosmic Rays*, 9.

11. Although Millikan didn't realize it at the time, this was actually the first indication of a latitude effect on cosmic rays, a phenomenon that was to cause him much consternation later.

12. Quoted in Kargon, *The Rise of Robert Millikan*, 138.

13. Quoted in Carlson, "A Century of Cosmic Rays," 32.

14. Kargon, *The Rise of Robert Millikan*, 28.

15. Kargon, 58.

16. Kargon, 59.

17. Judith R. Goodstein, *Millikan's School: A History of the California Institute of Technology* (New York: W. W. Norton & Co., 1991), 92. Goodstein, Judith R. *Millikan's School: A History of the California Institute of Technology* (New York: W. W. Norton & Co., 1991).

18. *The Electron: Its Isolation and Measurement and the Determination of Some of Its Properties* was first published in 1917 and revised or republished several times afterward.

19. Robert Andrews Millikan, *The Electron: Its Isolation and Measurement and the Determination of Some of Its Properties*, ed. and introduction by Jesse W. M. DuMond (Chicago: University of Chicago Press, [1917] 1968), 83.

20. Quoted in David Goodstein, "In Defense of Robert Millikan," *Engineering & Science* no. 4 (2000): 31.

21. Quoted in Michelangelo De Maria, M. G. Ianniello, and Arturo Russo, "The Discovery of Cosmic Rays: Rivalries and Controversies between Europe and the United States," *Historical Studies in the Physical and Biological Sciences* 22, no. 1 (1991): 171.

22. Robert Andrews Millikan, "Alleged Sins of Science," in *Science and the New Civilization* (New York: Charles Scribner's Sons, 1930), 81.

23. R. A. Millikan and G. Harvey Cameron, "High Frequency Rays of Cosmic Origin III. Measurements in Snow-Fed Lakes at High Altitudes," *Physical Review* 28 (1926): 851, 856, https://doi.org/10.1103/PhysRev.28.851.

CHAPTER 4

1. Harvey Brace Lemon, *Cosmic Rays Thus Far* (New York: W. W. Norton & Co., 1936), 67.

2. Kargon, *The Rise of Robert Millikan*, 139.

3. "Millikan Rays," *New York Times*, November 12, 1925, 24.

4. "Millikan Rays," 24.

5. "Millikan Rays," 24.

6. Walter, "From the Discovery of Radioactivity," 29.

7. Walter, 28–29.

8. Quoted in De Maria, Ianniello, and Russo, "Discovery of Cosmic Rays," 178, 181.

9. Quoted in Kargon, *The Rise of Robert Millikan*, 140.

10. De Maria, Ianniello, and Russo, "Discovery of Cosmic Rays," 181.

11. Kargon, *The Rise of Robert Millikan*, 140.

12. Peter Galison, *How Experiments End* (Chicago: University of Chicago Press, 1987), 81.

13. "Millikan Confirms Cosmic Ray Exists," *New York Times*, October 29, 1926.

14. Quoted in Kargon, *The Rise of Robert Millikan*, 141.

15. Quoted in Kargon, 141.

16. "Earth Still Being Created, Mystery Rays Show," *Popular Mechanics* (June 1928): 927.

17. "The Chief" nickname: see Goodstein, *Millikan's School*, among others.

18. Kargon, *The Rise of Robert Millikan*, 144.

19. Galison, *How Experiments End*, 87–88.

20. "Creation Continues, Millikan's Theory," *New York Times*, March 18, 1928.

21. Waldemar Kaempffert, "Super X-Rays Reveal the Secret of Creation," *New York Times*, March 25, 1928.

22. Kaempffert, "Super X-Rays Reveal the Secret of Creation."

23. "A World Center of Education and Research," *Los Angeles Times*, January 2, 1929, 109.

24. "A World Center of Education and Research," 109.

25. Lee Shippey, "The Lee Side of L.A.," *Los Angeles Times*, July 21, 1929, 18.

26. Shippey, "The Lee Side of L.A.," 18.

27. Associated Press, November 21, 1929.

28. One of the many inaccurate references of the time giving Millikan undeserved credit.

29. James Warnack, "Sermons on Literature," *Los Angeles Times*, April 28, 1929, 55.

30. "Faith to Move Mountains," *Los Angeles Times*, November 24, 1929, 20.

31. Quoted in Alex Grand, "History of Cosmic Rays in Comic Books," May 21, 2018, https://comicbookhistorians.com/cosmic-rays-in-comics-by-alex-grand/.

32. Kargon, *The Rise of Robert Millikan*, 143–144.

33. Although he didn't quite realize it at the time, Rossi had invented something that would later become a fundamental element of computer science, called the AND gate, a logic circuit that produces output only when all inputs are the same.

34. Dimitry V. Skobeltzyn, "The Early Stage of Cosmic Ray Particle Research," in Sekido and Elliot, eds., *Early History of Cosmic Ray Studies*, 47.

35. Skobeltzyn, "The Early Stage of Cosmic Ray Particle Research," 48.

CHAPTER 5

1. "'Tech Murder Mysteries' Out," *Los Angeles Times*, May 23, 1930, 28.

2. "'Tech Murder Mysteries' Out," *Los Angeles Times*, May 23, 1930, 28.

3. "'Tech Murder Mysteries' Out," *Los Angeles Times*, May 23, 1930, 28.

4. Jesse W. M. DuMond, "Editor's Introduction" in Millikan, *The Electron: Its Isolation and Measurement and the Determination of Some of Its Properties* (Chicago: University of Chicago Press, [1917] 1963), xli.

5. "Millikan Talks on Cosmic Rays," *Los Angeles Times*, June 6, 1930, 14.

6. "Dr. Millikan Goes Talkie at Caltech," *Los Angeles Times*, June 7, 1930, 7.

7. "Earth Committing Suicide?," *Los Angeles Times*, July 27, 1930, 117.

8. Compton, "Personal Reminiscences," 10.

9. Compton, 11.

10. Compton, 16.

11. Samuel K. Allison, *Arthur Holly Compton 1892–1962: A Biographical Memoir* (Washington, DC: National Academy of Science, 1965), 85.

12. Compton, "Personal Reminiscences," 36.

13. Compton, 41–42.

14. Compton, 41–42.

15. Compton, 41–42.

16. Michelangelo De Maria and Arturo Russo, "Cosmic Ray Romancing: The Discovery of the Latitude Effect and the Compton-Millikan Controversy," *Historical Studies in the Physical and Biological Sciences* 19, no. 2 (1989): 228, https://doi.org/10.2307/27757626.

17. Reminiscences of William Leonard Laurence (1964), Oral History Archives at Columbia, Rare Book and Manuscript Library, Columbia University, New York, 190–191.

18. Bailey Millard, "The Immortal Atom," *Los Angeles Times*, July 3, 1930, 22.

19. Millard, "The Immortal Atom," 22.

20. William Laurence, "Studies of the Cosmic Ray Point to Endless Creation," *New York Times*, September 28, 1930, XX4.

21. "Millikan En Route to Far North," *Los Angeles Times*, August 17, 1930, 13.

22. "Millikan Adds to Data on Ray," *Los Angeles Times*, September 14, 1930, 13.

23. "Millikan Airs Wonders," *Los Angeles Times*, September 23, 1930, 23.

24. "Dr. Millikan Has Bronchitis," *Los Angeles Times*, November 11, 1930, 8.

25. "Says Science Is Not Cause of Unemployment," *Chicago Tribune*, December 13, 1930, 9.

26. "Einstein Wish to Be Heeded," *Los Angeles Times*, December 29, 1930, 15.

CHAPTER 6

1. "Scientist Lounges While Dictating Notes," *Associated Press*, December 28, 1930.

2. "Text of Millikan's Address on Origin and Destiny of Matter," *New York Times*, December 30, 1930.

3. "Text of Millikan's Address on Origin and Destiny of Matter."

4. "Text of Millikan's Address on Origin and Destiny of Matter."

5. "Text of Millikan's Address on Origin and Destiny of Matter."

6. "Evolution Theory Is Not Atheistic, Declares Millikan," *Philadelphia Inquirer*, December 30, 1930, 1.

7. Reminiscences of William Leonard Laurence (1964), Oral History Archives at Columbia, 192.

8. William L. Laurence, "Millikan Finds Creation Still Goes On While Creator Directs the Universe," *New York Times*, December 30, 1930, 1.

9. "The Cosmic Ray and the Psalmist," *Philadelphia Inquirer*, January 1, 1931, 8.

10. Reminiscences of William Leonard Laurence (1964), Oral History Archives at Columbia, 193.

11. "Prof. Einstein Reaches Haven," *Los Angeles Times*, January 1, 1931, 2.

12. "Einstein Wish to be Heeded," 15.

13. Goodstein, *Millikan's School*, 100.

14. Quoted in Goodstein, 101.
15. "Full Aid Offered Einstein," *Los Angeles Times*, January 8, 1931, 19.
16. "Fellow Sages Honor Einstein," *Los Angeles Times*, January 16, 1931, 20.
17. "Savant Rivals in Agreement," *Los Angeles Times*, April 29, 1931, 28.
18. "Sky-Age Key Eludes Jeans," *Los Angeles Times*, May 7, 1931, 23.
19. "Famous Men of Science Meet," *Boston Globe*, May 8, 1931, 13.
20. "British Astronomer Confirmed in Belief," *Los Angeles Times*, May 10, 1931, 20.
21. "British Astronomer Confirmed in Belief," 20.
22. "Michelson Succumbs," *Los Angeles Times*, May 10, 1931, 1.
23. "Universe Held Mere Bubble," in *Los Angeles Times*, September 30, 1931, 3.
24. "Millikan vs. Jeans," *Los Angeles Times*, October 3, 1931, 18.
25. Auger, *What Are Cosmic Rays?*, 14–15.
26. David H. DeVorkin, *Race to the Stratosphere: Manned Scientific Ballooning in America* (New York: Springer-Verlag, 1989), 22.
27. "Little Information Expected from Ascent," *Los Angeles Times*, May 28, 1931, 2.
28. "The Constancy of Cosmic Rays," *Physical Review* 38 (1931): 1566.
29. "'Oh, Professor, Tell Us,'" *Denver Post*, September 20, 1931.
30. Charles Weiner, oral history interview of Betty Compton, April 11, 1968, Niels Bohr Library & Archives, American Institute of Physics, College Park, MD, https://www.aip.org/history-programs/niels-bohr-library/oral-histories/4560-1.
31. Quoted in De Maria and Russo, "Cosmic Ray Romancing," 224.
32. Quoted in De Maria and Russo, 225.
33. De Maria and Russo, 225.

CHAPTER 7

1. "Compton Will Hunt Secret of Cosmic Ray in World-Wide Study from Lofty Mountains," *New York Times*, January 3, 1932, 1.
2. "Science Finds Cosmic Clew to Human Destiny," *Chicago Tribune*, January 3, 1932, 1.
3. "Tailor-Made Souls," *Boston Globe*, January 6, 1932, 16.
4. "Tailor-Made Souls," 16.
5. Arthur Holly Compton, "A World Survey of Cosmic Rays," in Johnston, ed., *The Cosmos of Arthur Holly Compton*, 161.
6. Many notes in Compton's data sheets and notebooks reference "the bomb," which at first can be rather startling to a modern reader.
7. John J. Compton, "Adventures of a Citizen Scientist," *Perspectives on Science and Christian Faith* 62, no. 1 (March 2010), 56.

8. “The Origin of the Cosmic Ray: Science Attacks Basic Problem,” *New York Times*, January 10, 1932.

9. “The Origin of the Cosmic Ray: Science Attacks Basic Problem.”

10. Weiner, oral history interview of Betty Compton.

11. H. Victor Neher, “Some Reminiscences of the Early Days of Cosmic Rays,” in *The Birth of Particle Physics*, ed. Laurie M. Brown and Lillian Hoddeson (Cambridge: Cambridge University Press, 1983), 120–121.

12. “Monastery of science” quoted in DuMond, introduction to Millikan, *The Electron*, xlii.

13. Neher, “Some Reminiscences of the Early Days,” 122.

14. Neher, 123.

15. Neher, 123.

16. Arnold would go on to become the five-star Commanding General of the Army Air Forces in World War II and then a founding figure of the U.S. Air Force. He remained a strong advocate of scientific and technological innovation for his entire career.

17. Robert Millikan to Lt. Colonel Henry Arnold, July 6, 1932, California Institute of Technology Archives, Robert Andrews Millikan (RAM) Papers, 22.2.

18. Lt. Colonel Henry Arnold to Robert Millikan, July 9, 1932, RAM Papers, 22.2.

19. Robert Millikan to Lt. Colonel Henry Arnold, August 10, 1932, RAM Papers, 22.3.

20. Neher, “Some Reminiscences of the Early Days,” 124.

21. Neher, 124.

22. Neher, 125.

23. Arthur Compton to Robert Millikan, February 14, 1932, RAM Papers, 22.2.

24. Nikola Tesla, “The Cosmic Rays,” *New York Times*, February 6, 1932.

25. Michael Pupin to Arthur Compton, February 8, 1932, RAM Papers, 22.2.

26. Arthur Compton to Robert Millikan, February 14, 1932, RAM Papers, 22.2.

CHAPTER 8

1. “Starts Around World to Measure Cosmic Rays,” *Chicago Tribune*, March 19, 1932, 10; Bailey Millard, “Cosmic Rays Call Scientist,” *Los Angeles Times*, March 20, 1932, 27; “Mystery Ray Quest Begins,” *Los Angeles Times*, March 27, 1932, 23; “Cosmic Ray Quest over World Starts,” *Philadelphia Inquirer*, March 27, 1932, 2.

2. “Dr. Compton Gives Final Test to His ‘Cosmic Ray Trap,’” *Chicago Tribune*, March 16, 1932, 5; “To Test Force of Cosmic Rays,” *Boston Globe*, March 16, 1932, 9.

3. This of course would be news to their actual credited discoverer, the beleaguered Victor Hess.

4. “University of Chicago” was and is the actual name of the institution. It is unclear whether misnaming the University was an honest mistake by the editors or a sly insult.

5. "New Cosmic Ray Study," *Los Angeles Times*, January 18, 1932, 16.

6. "A Cosmic Ray Hunt," *Los Angeles Times*, March 21, 1932, 18.

7. "Compton Sees a New Epoch in Science," *New York Times Magazine*, March 13, 1932, 6.

8. Charles Weiner, oral history interview of Carl Anderson, June 30, 1966, Niels Bohr Library & Archives, American Institute of Physics, College Park, MD, https://www.aip.org/history-programs/niels-bohr-library/oral-histories/4487.

9. Carl D. Anderson, "Unraveling the Particle Content of the Cosmic Rays," in Sekido and Elliot, eds., *Early History of Cosmic Ray Studies*, 119.

10. Daniel J. Kevles, *The Physicists: The History of a Scientific Community in Modern America* (Cambridge, MA: Harvard University Press, 1995), 231.

11. Anderson, "Unraveling the Particle Content," 119.

12. Anderson, 120.

13. Anderson, 120.

14. Anderson, 123.

15. Weiner, oral history interview of Betty Compton.

16. Carpe expedition: "Cosmic Ray Expert Sails for Alaska," Associated Press, April 17, 1932; "Plane Helps in Mountain Climb," Associated Press, May 4, 1932; "Ray Quest Kills Pair," *Los Angeles Times*, May 17, 1932, 1; "Two Scientists Die Chasing Cosmic Rays in Alaska," *Chicago Tribune*, May 17, 1932, 16; Albert D. Lindley, "Mt. McKinley Tragedy Told," *Los Angeles Times*, May 19, 1932, 4; "Plane Missing on Alaskan Rescue Flight," *Chicago Tribune*, May 20, 1932, 7; "Two Scientists and Lost Pilot Safe on Glacier," *Chicago Tribune*, May 21, 1932, 3; "2 Scientists Will Remain on Alaska Glacier," *Chicago Tribune*, May 22, 1932, 17; "M.I.T. Man to Go to Mt. McKinley," *Boston Globe*, June 2, 1932, 19; "Dr. Bennett of Tech to Climb Mt. McKinley," *Boston Globe*, June 3, 1932, 16; "Find Explorer's Body on Glacier of Mt. McKinley," *Chicago Tribune*, August 22, 1932, 1.

17. Ransome Sutton, "Ray Data Hint Space Static," *Los Angeles Times*, May 22, 1932, 21.

18. Sutton, "Ray Data Hint Space Static," 21.

19. Sutton, "Ray Data Hint Space Static," 21.

20. "Cosmic Ray's Funny Phase in Versy Maze," *Los Angeles Times*, May 27, 1932, 26.

21. Arthur Compton, "Variation of the Cosmic Rays with Latitude," *Physical Review* 41 (1932), 111–113.

22. Compton, "Variation of the Cosmic Rays with Latitude," 111–113.

23. Compton, "Variation of the Cosmic Rays with Latitude," 111–113.

24. "New Cosmic Ray Penetrates 18 Feet of Lead," *Chicago Tribune*, June 25, 1932, 5.

25. *Chicago Tribune*, June 27, 1932, 12.

26. "New Data Challenge Cosmic Ray Theory," *New York Times*, July 8, 1932, 19.

27. "Airplanes and Balloons Will Dare Upper Reaches in New Cosmic Ray Test," *Los Angeles Times*, July 9, 1932, 8.

28. "Cosmic Rays Get Stronger at High Levels: Compton," *Chicago Tribune*, July 29, 1932, 12; "Millikan's Ray Theory Rapped," *Los Angeles Times*, July 29, 1932, 13; "Compton Confirms Cosmic Ray Beliefs," *New York Times*, July 29, 1932.

29. Bailey Millard, "Are Cosmic Rays Merely Electrons?," *Los Angeles Times*, August 4, 1932, 22.

30. "Benefits Seen in Machine Age," *Los Angeles Times*, July 29, 1932, 27.

CHAPTER 9

1. Georg Pfotzer, "On Erich Regener's Cosmic Ray Work in Stuttgart and Related Subjects," in Sekido and Elliot, eds., *Early History of Cosmic Ray Studies*, 88.

2. "Balloon Climbs 17 Miles and Cosmic Ray Theory Gets a Jolt," *Chicago Tribune*, August 14, 1932, 1; "Flight Assails Theory of Rays," *Los Angeles Times*, August 14, 1932, 3.

3. "Scales High Peak for Cosmic Rays," *Philadelphia Inquirer*, August 14, 1932, 6.

4. "Ray Circus," *Time*, August 29, 1932, 22.

5. On Millikan expeditions: "New Trap Made for Cosmic Ray," *Los Angeles Times*, August 19, 1932, 19; "Dr. Millikan Has Airplane Device to Record Ray," *Chicago Tribune*, August 19, 1932, 15; "Planes Test Cosmic Ray," *Los Angeles Times*, August 31, 1932, 17; "Millikan Goes on Ray Quest," *Los Angeles Times*, September 2, 1932, 20; "Millikan Starts for Arctic," *New York Times*, September 3, 1932.

6. "Compton Gives New Evidence on Cosmic Ray," *Chicago Tribune*, September 3, 1932, 6; "Eclipse Study Reveals Cosmic Rays as Electrons," *Boston Globe*, September 3, 1932, 13; "Rays Indicated as Electrons," *Los Angeles Times*, September 3, 1932, 3; "Eclipse Observer Finds Cosmic Rays Resemble X-Ray," *Philadelphia Inquirer*, September 3, 1932, 2.

7. "Cosmic Rays Photograph Selves by British Device," *Los Angeles Times*, September 3, 1932, 3; "Photographs Itself," *Chicago Tribune*, September 3, 1932, 6.

8. "A Raid on the Rays," *Los Angeles Times*, September 6, 1932, 18; "Millikan Ready for Rays Test," *Los Angeles Times*, September 7, 1932, 1; "Millikan to Soar 25,000 Feet Today for Cosmic Rays," *Philadelphia Inquirer*, September 7, 1932, 2.

9. "Airmen Back from Cosmic Ray Flight," *Boston Globe*, September 8, 1932, 7; "Returns Bearing Millikan Apparatus," *New York Times*, September 8, 1932; "Completes 2d Skyscraping Flight for Millikan," *Chicago Tribune*, September 9, 1932, 15.

10. "Sees Universe as Expanding," *Boston Globe*, September 8, 1932, 7; "Eddington Depicts Growing Universe as Dilating Sphere," *Philadelphia Inquirer*, September 8, 1932, 1.

11. "Sees Universe as Expanding," 7; "Eddington Depicts Growing Universe as Dilating Sphere," 1.

12. "Millikan Plans Winnipeg Visit," *Los Angeles Times*, September 10, 1932, 1; "Millikan's Ray Quest to Shift," *Los Angeles Times*, September 11, 1932, 1.

13. "Millikan Upholds Religious Belief," *New York Times*, September 12, 1932; "Old Dogmas Death Seen by Millikan," *Los Angeles Times*, September 12, 1932, 3; "Dr. Millikan Pictures Science and Religion Near Common Goal," *Philadelphia Inquirer*, September 12, 1932, 3.

14. "Dakota Experiment Balked by Winds," *Los Angeles Times*, September 15, 1932, 3; "Millikan Test Balloon Falls," *Los Angeles Times*, September 17, 1932, 1.

15. "Cosmic Ray Results to be Delayed," *Los Angeles Times*, September 20, 1932, 7; "Dr. Millikan's Results Must First Be Compiled," *Boston Globe*, September 20, 1932, 9; "Millikan Back from Ray Test," *Los Angeles Times*, September 30, 1932, 19.

CHAPTER 10

1. "Brings Arctic Data on the Cosmic Ray," *New York Times*, September 15, 1932; "Millikan's Ray Theory Denied," *Los Angeles Times*, September 15, 1932, 3; "Compton Back from Arctic, Finds Cosmic Ray Is Electrical," *Chicago Tribune*, September 15, 1932, 12; "More Cosmic Rays at Magnetic Pole," *Boston Globe*, September 16, 1932, 32.

2. "Terrific Atomic Blasts Discovered on Mountain," *Los Angeles Times*, September 18, 1932, 1; "Added Data on Ray Dispute Millikan," *New York Times*, September 20, 1932.

3. "Compton Describes Cosmic Ray Findings," *New York Times*, September 17, 1932.

4. Frank B. Jewett to Robert Millikan, September 28, 1932, RAM Papers, 40.25.

5. "Cosmic-Ray Romancing," *New York Times*, September 18, 1932.

6. "It Is Done with Mathematics," *New York Times*, September 18, 1932.

7. "How a Newspaper Looks at Cosmic Rays," *Science* 76, no. 1969 (September 23, 1932): 276.

8. "The Ultimate Particle," *Los Angeles Times*, September 18, 1932, 28.

9. "Uppish," *Los Angeles Times*, September 27, 1932, 20.

10. W. I. (Fox) Stanton, "Football Success Recipe Given," *Los Angeles Times*, October 2, 1932, 68.

11. Neher, "Some Reminiscences of the Early Days," 125.

12. "Quotations: Professor Arthur H. Compton's Studies of Cosmic Rays," *Science* 76, no. 1971 (October 7, 1932): 325.

13. "Compton Is Back with New Facts on Cosmic Ray," *Chicago Tribune*, October 2, 1932, 16; "Compton Declares Cosmic Ray Electric," *New York Times*, October 3, 1932; "Cosmic Ray Called Vagrant of Space," *Philadelphia Inquirer*, October 3, 1932, 2.

14. "Scientists at Loggerheads," *Los Angeles Times*, October 2, 1932, 111.

15. "Home Cosmic Rays Trapped," *Los Angeles Times*, October 9, 1932, 60; "Down to Business," *Los Angeles Times*, October 13, 1932, 20.

16. Jesse W. M. DuMond, introduction to Millikan, *The Electron*, xli.

17. "Medal Given Millikan, 'Prophet of New Time,'" *Los Angeles Times*, October 28, 1932, 3; "Millikan Hailed as Prophet of New Time and Given Medal," *Chicago Tribune*, October 28, 1932, 19; "Millikan Receives Roosevelt Medal," *New York Times*, October 28, 1932; "More Honors for Millikan," *Los Angeles Times*, November 3, 1932, 20.

18. "Town Hall Lectures," *Chicago Tribune*, October 9, 1932, 62.

19. "Compton to Tell of His Hunt for Cosmic Ray Data," *Chicago Tribune*, October 16, 1932, 4; "Compton Turns More Light on Cosmic Ray," *Chicago Tribune*, October 20, 1932, 13; "Tells of Rays," *Chicago Tribune*, October 26, 1932, 7.

20. Ruth De Young, "Dr. Millikan Sees Hope of Eternal Earth," *Chicago Tribune*, October 21, 1932, 24.

21. "All About Cosmic Rays," *Chicago Tribune*, October 30, 1932, 62.

22. "Electroscope Returned," *Los Angeles Times*, October 22, 1932, 9.

23. "Guard Flyer to Test Cosmic Ray Here Today," *Los Angeles Times*, October 27, 1932, 17; "New Ray Test Made above City," *Los Angeles Times*, October 28, 1932, 24; "Savant Goes to Andes for Cosmic Ray Tests," *Los Angeles Times*, October 29, 1932, 16.

24. "A Prophet of the New Time," *New York Times*, October 29, 1932.

25. "Millikan Receives Roosevelt Medal," *New York Times*, October 28, 1932.

26. "More Honors for Millikan," *Los Angeles Times*, November 3, 1932, 20.

27. "Medal Given Millikan, 'Prophet of New Time,'" *Los Angeles Times*, October 28, 1932, 3.

28. Bob Ray, "Bulldogs Face Caltech Next," *Los Angeles Times*, November 2, 1932, 27.

29. Waldemar Kaempffert, "The Cosmic Riddle That Awes Science," *New York Times*, November 13, 1932.

30. "Pictures Cosmic Ray Born 100 Miles Up," *New York Times*, November 15, 1932; "Cosmic Rays Believed to Come from the Upper Atmosphere," *Boston Globe*, November 15, 1932, 10; "Rays Linked to Earth," *Los Angeles Times*, November 15, 1932, 1; Waldemar Kaempfert, "Cosmic Rays Called Product of Stratosphere," *Chicago Tribune*, November 15, 1932, 16.

31. "Scientists Work Hard, Fact Chases Theory," *New York Times*, November 16, 1932.

32. "Compton on Cosmic Rays," *Los Angeles Times*, November 20, 1932, 24.

CHAPTER 11

1. "Wanted—2 Yanks for Flight to Stratosphere," *Chicago Tribune*, November 25, 1932, 10; "Piccard Plans Two Ascensions," *Boston Globe*, December 21, 1932, 14.

2. Neher, "Some Reminiscences of the Early Days," 125–126.

3. Neher, "Some Reminiscences of the Early Days," 125–126.

4. Neher, "Some Reminiscences of the Early Days," 125–126.

5. The *Los Angeles Times* called Compton a "friendly" rival, while the New York report spared that description.

6. "Millikan Offers Cosmic Ray Facts," *New York Times*, December 3, 1932; "Answer Given on Cosmic Ray," *Los Angeles Times*, December 3, 1932, 13; "Millikan Clings to Photon Theory," *New York Times*, December 4, 1932.

7. "Millikan Fires Back," *Los Angeles Times*, December 6, 1932, 20.

8. De Maria and Russo, "Cosmic Ray Romancing," 242.

9. Ransome Sutton, "Abbe Expounds Cosmos Theory," *Los Angeles Times*, December 4, 1932, 25; "Cosmic Theories Argued," *Los Angeles Times*, December 8, 1932, 32.

10. "Sun Is Not Source of Cosmic Radiation," *Philadelphia Inquirer*, December 17, 1932, 4; "Stars' Energy Gives Us Pep," *Los Angeles Times*, December 18, 1932, 58.

11. "Cosmic Rays Ring Bell for Lecturer," *Boston Globe*, December 12, 1932, 17.

12. "Cosmic Ray Survey Finished at MIT," 8.

13. "What's New in Science: From the Academicians," *Los Angeles Times*, December 25, 1932, 87.

14. AAAS meeting details: "Symposia at Atlantic City," *Science* 76, no. 1976 (November 11, 1932): 440–441; "Mathematics, Physics and Astronomy at Atlantic City," *Science* 76, no. 1979 (December 2, 1932): 517–518; "The Science Exhibition," *Science* 76, no. 1982 (December 23, 1932): 601; "The Atlantic City Meeting," *Science* 76, no. 1982 (December 23, 1932): 599–600.

15. "Cosmic Ray Rivals to Meet in Debate," *New York Times*, December 26, 1932.

16. "Debate Is Ahead on Cosmic Rays," *Boston Globe*, December 26, 1932, 2.

17. "Cosmic and Comic Rays," *Boston Globe*, December 27, 1932, 16.

18. "Cosmic and Comic Rays," 12.

19. Robert Millikan to Arthur Compton, November 30, 1932, Arthur Holly Compton (AHC) Papers, University Archives, Washington University in St. Louis, series 2, box 4, folder 22.

20. Arthur Compton to Robert Millikan, December 5, 1932, AHC Papers, series 2, box 4, folder 22.

21. "Mathematics, Physics and Astronomy at Atlantic City," *Science* 76, no. 1979 (December 2, 1932): 517–518.

22. "Millikan Retorts Hotly to Compton in Cosmic Ray Clash," *New York Times*, December 31, 1932.

23. Neher, "Some Reminiscences of the Early Days," 126–127.

24. Kevles, *The Physicists*, 242.

25. Neher, "Some Reminiscences of the Early Days," 127.

CHAPTER 12

1. Compton, "A Geographic Study of Cosmic Rays," *Physical Review* 43, no. 6 (March 15, 1933): 387–403.

2. Quoted in Kevles, *The Physicists*, 241.

3. "New Techniques in the Cosmic-Ray Field and Some of the Results Obtained with Them," *Physical Review* 43 (April 15, 1933): 661.

4. De Maria and Russo, "Cosmic Ray Romancing," 246.

5. "New Techniques in the Cosmic-Ray Field and Some of the Results Obtained with Them," *Physical Review* 43 (April 15, 1933): 666.

6. Compton and Millikan: William L. Laurence, "Millikan Retorts Hotly to Compton in Cosmic Ray Clash," *New York Times*, December 31, 1932, 1; "Dr. Compton Finds Cosmic Ray Varies," December 31, 1932, 6; "Millikan Reports Views Unaltered," December 31, 1932, 6; "Millikan and Compton Debate Cosmic Ray Facts and Theories," *Science Newsletter* 23, no. 613, January 7, 1933, 6–7.

7. Laurence, "Millikan Retorts Hotly to Compton in Cosmic Ray Clash," 1; "Dr. Compton Finds Cosmic Ray Varies," 6; "Millikan Reports Views Unaltered," 6; "Millikan and Compton Debate Cosmic Ray Facts and Theories," 6–7.

8. "Cosmic Mysteries Still Mysterious," *Philadelphia Inquirer*, December 31, 1932, 7; "Cosmic Rays Still Elusive," *Los Angeles Times*, December 31, 1932, 3; "AAAS at Atlantic City," *TIME*, January 9, 1933, 34–35.

9. "The Cosmic Ray Riddle," *New York Times*, December 31, 1932, 14.

10. "Nobel Winners Debate on Rays," *Philadelphia Inquirer*, December 31, 1932, 1.

11. Reminiscences of William Leonard Laurence (1964), Oral History Archives at Columbia, 195–196.

12. Telegram from Millikan to Editor, *New York Times*, January 1, 1933, RAM Papers, 22.3; "Millikan Denies 'Clash' on Theory," *New York Times*, January 1, 1933, 16.

13. "AAAS at Atlantic City," *TIME*, January 9, 1933, 34–35.

14. Reminiscences of William Leonard Laurence (1964), Oral History Archives at Columbia, 196.

15. "Dr. Millikan's Statement Evokes Reply from Writer," *New York Times*, January 8, 1933, E5.

16. Reminiscences of William Leonard Laurence (1964), 196.

17. "The Week in Science: The Debate Concerning the Cosmic Rays," *New York Times*, January 1, 1933, XX4.

CHAPTER 13

1. "Einstein in U.S. with His Celluloid Collar," *Boston Globe*, January 10, 1933, 8; "Einstein Guest of Southland," *Los Angeles Times*, January 10, 1933, 15.

2. "Cosmic Ray Photographs Stolen from Compton's Aid," *Chicago Tribune*, February 4, 1933, 7.

3. "Sends Balloons up 17 Miles on Cosmic Ray Trip," *Chicago Tribune*, January 5, 1933, 1.

4. "Piccard Discusses Cosmic Ray Theory," *Philadelphia Inquirer*, January 14, 1933, 3; "Piccard Sees New Power," *Los Angeles Times*, January 14, 1933, 3; "Piccard Sees Cosmic Rays as Future's Energy," *Chicago Tribune*, January 14, 1933, 1.

5. "Power from Space," *Los Angeles Times*, January 17, 1933, 18; "Little Drops of Water," *Los Angeles Times*, January 28, 1933, 16.

6. "A Line o' Type or Two," *Chicago Tribune*, January 16, 1933, 10.

7. James J. Montague, "Alibi," *Los Angeles Times*, January 24, 1933, 16.

8. "Creation Seen as Explosion," *Los Angeles Times*, January 12, 1933, 22; "Believes Cosmic Rays Universe's Birth Cry," *Boston Globe*, January 12, 1933, 4; "Abbe le Maitre Calls Cosmic Ray Echo of Universe's 'Birth Cry,'" *Philadelphia Inquirer*, January 12, 1933, 3; "Abbe Lemaitre on Cosmic Rays," *Science Service* 77, no. 1986 (January 20, 1933): 6a–8a; "Einstein Backs Lemaitre Idea That Cosmic Rays Are Birth Cries," *Science News Letter* 23, no. 615, January 21, 1933, 37; "Visiting Eminence," *TIME*, January 23, 1933, 30–31.

9. "Creation Seen as Explosion," 22; "Believes Cosmic Rays Universe's Birth Cry," 4; "Abbe le Maitre Calls Cosmic Ray Echo of Universe's 'Birth Cry,'" 3; "Abbe Lemaitre on Cosmic Rays," 6a–8a; "Einstein Backs Lemaitre Idea That Cosmic Rays Are Birth Cries," 37; "Visiting Eminence," 30–31.

10. "Creation Seen as Explosion," 22; "Believes Cosmic Rays Universe's Birth Cry," 4; "Abbe le Maitre Calls Cosmic Ray Echo of Universe's 'Birth Cry,'" 3; "Abbe Lemaitre on Cosmic Rays," 6a–8a; "Einstein Backs Lemaitre Idea That Cosmic Rays Are Birth Cries," 37; "Visiting Eminence," 30–31.

11. "Compton's Theory Checked by Figures," *New York Times*, January 28, 1933; Lemaitre and Vallarta, "On Compton's Latitude Effect of Cosmic Radiation," *Physical Review* 43, no. 87 (January 15, 1933).

12. "Millikan's Data Confirm Compton," *New York Times*, February 5, 1933; "Cosmic Rays Found Stronger at Poles," *Philadelphia Inquirer*, February 5, 1933, 4.

13. Millikan, "New Techniques . . . ," *Physical Review* 43, April 15, 1933, 667–668.

14. De Maria and Russo, "Cosmic Ray Romancing," 251, 254.

15. "Millikan to Compton," *TIME*, February 13, 1933, 38.

16. "Millikan's Data Confirm Compton," *New York Times*, February 5, 1933.

17. "Cosmic Rays to Be Heard on Radio Network," *New York Times*, February 9, 1933.

18. "Prof. Piccard to Lecture Here on Stratosphere," *Chicago Tribune*, February 8, 1933, 8; "Piccard Speaks Tomorrow Night on N.U. Campus," *Chicago Tribune*, February 14, 1933, 18.

19. DeVorkin, *Race to the Stratosphere*, 49.

20. Arthur Compton to Robert Millikan, May 9, 1933, RAM Papers, 22.4.

21. Robert Millikan to Arthur Compton, May 15, 1933, RAM Papers, 22.4.

22. DeVorkin, *Race to the Stratosphere*, 63–69.

23. Editorial, *Los Angeles Times*, April 2, 1933, 91.

24. "Millikan Links Hair Combing to Discussing Cosmic Rays," *Philadelphia Inquirer*, April 21, 1933, 28; "Cosmic Ray Source Not in Upper Air," *Los Angeles Times*, April 22, 1933, 3.

25. "Editorial Comment," *Philadelphia Inquirer*, April 26, 1933, 10.

26. "Science Grasps Cosmic Clews to All Creation," *Chicago Tribune*, March 26, 1933, 10.

27. "Fair to Explain Myriad Marvels of Electricity," *Chicago Tribune*, May 22, 1933, 5; "Star's Light to Brighten World's Fair," *Los Angeles Times*, May 22, 1933, 2.

28. "Science Builds a Temple of Its Wonders at Fair," *Chicago Tribune*, April 12, 1933, 28.

29. "Start Cleanup Today for Fair's Opening Throng," *Chicago Tribune*, May 24, 1933, 10.

30. "Cosmic Rays Converge," *Los Angeles Times*, May 22, 1933, 18.

31. "Prof. Compton in Crash; Wife and Son Are Injured," *Chicago Tribune*, May 12, 1933, 5.

32. "Flight into Stratosphere Scheduled," *Los Angeles Times*, April 28, 1933, 5; "Stratosphere Adventure Set," *Los Angeles Times*, May 23, 1933, 7.

33. Robert Millikan to Arthur Compton cable, May 24, 1933, RAM Papers, 22.4.

34. Arthur Compton to Robert Millikan, May 27, 1933, RAM Papers, 22.4.

35. "Balloon May Decide Dispute of Professors," *Boston Globe*, May 24, 1933, 13.

36. Dempster MacMurphy, "Piccard Flight May End Compton-Millikan Debate on Cosmic Ray Properties," *Chicago Daily News* (undated), in RAM Papers, 22.4.

CHAPTER 14

1. Hal Foust, "A. P. Sloan Jr. Hails New Day in Fair Speech," *Chicago Tribune*, May 25, 1933, 5.

2. "Plane Climbs Six Miles with Cosmic Ray Trap," *Los Angeles Times*, June 8, 1933, 10.

3. Editorial, *Los Angeles Times*, June 9, 1933, 8.

4. "Atom May Yield 2 New Particles," *Philadelphia Inquirer*, June 19, 1933, 2; William L. Laurence, "Atom Study Backs Creation Theory," *New York Times*, June 22, 1933, 1; Philip Kinsley, "Millikan Sees Energy Flowing from the Stars," *Chicago Tribune*, June 22, 1933, 8; "Millikan Says Nature's at Work," *Los Angeles Times*, June 22, 1933, 9.

5. "Wellesley Institute to Hear Dr. Millikan," *Boston Globe*, June 13, 1933, 11.

6. Reminiscences of William Leonard Laurence (1964), Oral History Archives at Columbia, 196–197.

7. Robert Millikan to C. C. Mallory, July 14, 1933; Robert Millikan to Stanley Dollar, July 14, 1933, RAM Papers, 22.4.

8. A. W. Stevens to Robert Millikan, July 21, 1933, RAM Papers, 22.4.

9. A. W. Stevens to Robert Millikan, July 21, 1933, RAM Papers, 22.4.

10. A. W. Stevens to Robert Millikan, July 21, 1933, RAM Papers, 22.4.

11. A. W. Stevens to Robert Millikan, July 21, 1933, RAM Papers, 22.4.

12. Robert Millikan to A. W. Stevens, August 5, 1933, RAM Papers, 22.4.

13. Robert Millikan to A. W. Stevens, August 5, 1933, RAM Papers, 22.4.

14. "Byrd Party to Aid Cosmic Ray Study," *New York Times*, July 21, 1933; "Editorial Points," *Boston Globe*, July 22, 1933, 14.

15. Settle flight preparations: "Stratosphere Ball Tested," *Boston Globe*, July 8, 1933, 20; "Stratosphere Tests Made," *Los Angeles Times*, July 12, 1933, 9.

16. "Settle to Make Solo Trip into Stratosphere," *Chicago Tribune*, July 23, 1933, 2.

17. Arthur H. Compton, "Balloonists Seek Secret of Cosmic Rays," *Boston Globe*, July 7. 1933, 7; Arthur H. Compton, "Cosmic Ray Mystery May Be Solved Soon," *Boston Globe*, July 10, 1933, 11.

18. Robert A. Millikan, "Stratosphere Cosmic Ray Data Is Goal of Millikan," *Boston Globe*, July 11, 1933, 13.

19. "New Data Given on Cosmic Rays," *Los Angeles Times*, July 15, 1933, 3; "Bares Nature of Cosmic Ray," *Boston Globe*, July 15, 1933, 3; "Both Sides Declared Right in Cosmic Ray Arguments," *Philadelphia Inquirer*, July 15, 1933, 24; "No Electrons in Cosmic Rays," *Chicago Tribune*, July 15, 1933, 6; Bailey Millard, "Mysterious Rays," *Los Angeles Times*, July 30, 1933, 16.

20. Settle flight: "U.S. Naval Aide Poised for Stratosphere Jaunt," *Philadelphia Inquirer*, August 5, 1933, 1; "Balloonist Set to Go," *Los Angeles Times*, August 5, 1933, 1; "Weather Best in a Month," *New York Times*, August 5, 1933; "Balloon Rises a Mile; Falls," *Chicago Tribune*, August 5, 1933, 1; "Settle Down," *TIME*, August 14, 1933.

21. Robert Millikan to Arthur Compton, August 7, 1933, RAM Papers, 22.4.

22. "Grace Skipper to Aid Science Research," *Los Angeles Times*, August 12, 1933, 4; "Electroscopes Record Rays on Ocean," *Los Angeles Times*, August 13, 1933, 52; "Cosmic Ray Studied on Round-World Liner; Millikan Device Expected to Bring New Data," *New York Times*, August 27, 1933.

23. C. C. Mallory to Robert Millikan, August 3, 1933, RAM Papers, 22.4.

CHAPTER 15

1. Arthur Compton to Robert Millikan, August 12, 1933, RAM Papers, 22.4.

2. The record would eventually be disallowed ostensibly because the Soviet Union was not a member of the FAI at the time of the flight, although political considerations likely also played a factor.

3. "Russians Claim Balloon Mark of 11 Miles High," *Chicago Tribune*, October 1, 1933, 6.

4. "Russians Claim Balloon Mark of 11 Miles High," 6.

5. Arthur Compton to Robert Millikan, October 26, 1933, RAM Papers, 22.5.

6. Robert Millikan to Arthur Compton, October 31, 1933, RAM Papers, 22.5.

7. For example, "Intensity of Cosmic Rays in Boston Put on Record," *Boston Globe*, October 30, 1933, 3; "Paths of Science Study Ships to Cross," *Los Angeles Times*,

November 19, 1933, 25; "Electroscopes Arrive on Ships," *Los Angeles Times*, November 20, 1933, 17.

8. Back in August, Fordney had led the Marines that salvaged the balloon from the Chicago rail yards.

9. Settle-Fordney flight: "Up 53,000 Feet in Latest Report," *Boston Globe*, November 20, 1933, 1; "Soar 11 Miles, 'Lost' Landing," *Boston Globe*, November 21, 1933, 1; "American Balloon Soars More than 58,000 Feet," *Los Angeles Times*, November 21, 1933, 1; "Settle Balloon Believed Down in Jersey Marsh," *Philadelphia Inquirer*, November 21, 1933, 1; "Stratosphere Pair Safe in New Jersey," *Boston Globe*, November 21, 1933, 1; "59,000 Mark Topped by Settle in 500-Mile Drift," *Philadelphia Inquirer*, November 22, 1933, 1.

10. "In the Stratosphere," *Boston Globe*, November 22, 1933, 16.

11. "Into the Stratosphere," *Los Angeles Times*, November 22, 1933, 20.

12. "Finds Cosmic Rays More Numerous in Stratosphere," *Chicago Tribune*, December 9, 1933, 12.

13. Editorial, *Boston Globe*, December 11, 1933, 14.

14. "Will Show New Gains by Science," *Boston Globe*, December 27, 1933, 7.

15. "All Matter from Light," *Boston Globe*, December 29, 1933, 1.

16. Philip Kinsley, "Compton Finds Deity Is Basis of All Science," *Chicago Tribune*, December 26, 1933, 7; "'Attitude of Faith in God May Be Thoroughly Scientific,'" *Boston Globe*, December 26, 1933, 14, "Calls Faith in God a Scientific Stand," *Philadelphia Inquirer*, December 27, 1933, 26.

17. "Atomic Mist Found in Stratosphere," *Philadelphia Inquirer*, February 25, 1934, 2; "Explanation on Origin of Cosmic Rays Is Nearer," *Chicago Tribune*, February 25, 1934, 11; "Cosmic Ray Key Found," *Los Angeles Times*, February 25, 1934, 5; "Mystery of the Deathless Rays," *Philadelphia Inquirer*, February 26, 1934, 8; Bailey Millard, "Compton's Discovery," *Los Angeles Times*, March 4, 1934, 16.

18. William L. Laurence, "New Data on Rays Given by Millikan," *New York Times*, June 24, 1934; "Millikan Says Flight Proves Cosmic Ray Idea," *Chicago Tribune*, June 24, 1934, 21; "Cosmic Ray Test Told," *Los Angeles Times*, June 24, 1934, 1.

19. "Giant Balloon Soon to Soar," *Los Angeles Times*, May 7, 1934, 3; "World's Largest Balloon for Next Daring Flight," *Boston Globe*, June 6, 1934, 16.

20. Captain Albert Stevens to Robert Millikan, December 31, 1933, RAM Papers, 22.5.

21. "'Taffy Sticks' to Stick to Cosmic Rays in Stratosphere," *Boston Globe*, July 11, 1934, 11.

22. "Trips Aloft in Balloon Called Fun," *Los Angeles Times*, June 15, 1934, 10.

23. "Byrd Prepares Frigid Sortie," *Los Angeles Times*, June 28, 1934, 8.

24. "Balloon's Data Lost," *Los Angeles Times*, July 30, 1934, 2; "Cosmic Ray Film Not Smashed," *Boston Globe*, August 8, 1934, 15; "Cosmic Ray Film Saved from Flight Disaster," *Science News Letter*, October 13, 1934, 233.

25. "Stratosphere Trip Planned by Piccards," *Los Angeles Times*, August 21, 1934, 15.

26. Piccard flight: "Piccards Order Balloon Out," *Boston Globe*, October 23, 1934, 1; "Balloon Lands on Top of Tree," *Boston Globe*, October 24, 1934, 1; "Piccards Land, Soar 10 Miles in Stratosphere," *Chicago Tribune*, October 24, 1934, 1; "Piccard Bag Comes Down," *Los Angeles Times*, October 24, 1934, 2; "Piccards Suffer Minor Injuries in Landing of Balloon," *Chicago Tribune*, October 25, 1934, 3; "Prof. Piccard's Own Story of Stratosphere Flight," *Boston Globe*, October 25, 1934, 19.

27. "Gas Bag with Radio Brain to Signal Cosmic Ray Data," 10; "Globe-Girdling Traps Planned for Cosmic Ray," 3; "Balloon to Get Cosmic Ray Data Ready at U. of C.," 7; "Balloons Carrying Radios Tell Stratosphere Secrets," 131.

28. "Gas Bag with Radio Brain to Signal Cosmic Ray Data," *Boston Globe*, August 11, 1934, 10; "Globe-Girdling Traps Planned for Cosmic Ray," *Los Angeles Times*, August 11, 1934, 3; "Balloon to Get Cosmic Ray Data Ready at U. of C.," *Chicago Tribune*, August 11, 1934, 7; "Balloons Carrying Radios Tell Stratosphere Secrets," *Science News Letter*, September 1, 1934, 131.

29. "Ready to Send Up Tiny Balloon on Cosmic Ray Test," *Chicago Tribune*, August 29, 1934, 4; "Unoccupied Balloon Soared 17 3/4 Miles," *Boston Globe*, August 30, 1934, 4; "Dr. Compton's Tiny Balloon Rises 17 1/2 Miles," *Chicago Tribune*, August 30, 1934, 1.

30. "Next Radio Robot Balloon to Measure Cosmic Rays," *Science News Letter*, September 15, 1934, 163.

CHAPTER 16

1. Marcus Aurelius, *Meditations*, trans. Gregory Hays (New York: Modern Library, 2002), 4:49.

2. Ferdinand Kuhn Jr., "Millikan Modifies Creation Theory," *New York Times*, October 3, 1934; "Millikan Admits Destruction to Cosmic Theory," *Chicago Tribune*, October 3, 1934, 18; "Creation and Destruction," *TIME*, October 15, 1934, 44.

3. "New Cosmic Ray," *Los Angeles Times*, October 8, 1934, 24.

4. "The Week in Science: Origin of the Cosmic Rays," *New York Times*, October 21, 1934.

5. Cited in De Maria and Russo, "Cosmic Ray Romancing," 253 ref 108.

6. "Millikan Explains Modified Theory," *New York Times*, October 21, 1934; "Cosmic Rays," *Science News*, October 26, 1934, 6a–7a.

7. "Cosmic Ray to Make Bow," *Los Angeles Times*, December 24, 1934, 6; "Public to See Wonder Acts of 'Cosmic Rays,'" *Chicago Tribune*, December 24, 1934, 7.

8. Philip Kinsley, "Prof. Einstein Doubts Energy Theory's Value," *Chicago Tribune*, December 29, 1934, 2.

9. William L. Laurence, "Cosmic Ray Puzzle Due to Be Solved," *New York Times*, December 30, 1934; Philip Kinsley, "Millikan Sees Cosmic Rays as 'Superbandits,'" *Chicago Tribune*, December 30, 1934, 7; "Cosmic Ray 'Tall Stories' Debunked by

Dr. Millikan," *Philadelphia Inquirer*, December 30, 1934, 1; "Text of the Cosmic Ray Credo by Dr. Millikan," *New York Times*, December 30, 1934.

10. "What's New in the Progress of Science," *Los Angeles Times*, January 20, 1935, 121.

11. "Prof. Compton Home for New Strato Tests," *Chicago Tribune*, March 26, 1935, 4; Philip Kinsley, "Traces Sources of Cosmic Rays Past Milky Way," *Chicago Tribune*, April 14, 1935, 20.

12. Philip Kinsley, "Millikan Gives a New View of Science World," *Chicago Tribune*, March 16, 1935, 15.

13. L. G. H. Huxley, "Electrons," *Nature* 136, no. 3435 (1935): 320.

14. Edward Condon, "Recent Developments in Atomic Physics," *Review of Scientific Instruments* 6, no. 4 (1935): 90.

15. Kargon, *Rise of Robert Millikan*, 159.

16. Auguste Piccard would be so impressed by the feat that he would announce plans for yet another flight to surpass the record, but soon turned his attentions to deep sea exploring instead of skyscraping. The Piccard family would continue to break records, such as Auguste's grandson Bertrand, who would complete the first nonstop balloon trip around the world in 1999 and the first circumnavigation of the Earth in a solar-powered airplane in 2016.

17. "Strato Balloon up 74,000 Feet," *Boston Globe*, November 11, 1935, 1; "Record Heights Attained in Hop," *Boston Globe*, November 12, 1935, 1; "Tell of 14 Mile High Flight," *Chicago Tribune*, November 12, 1935, 1; "Stratosphere Flyers Soar Fourteen Miles," *Los Angeles Times*, November 12, 1935, 1.

18. "Data of Flyers May Aid Science," *Boston Globe*, November 12, 1935, 8.

19. Craig Ryan, *The Pre-Astronauts: Manned Ballooning on the Threshold of Space* (Annapolis, MD: Naval Institute Press, 1995), 60–61.

20. Ryan, *The Pre-Astronauts*, 61.

CHAPTER 17

1. Philip Kinsley, "Compton on Cosmic Rays," *Chicago Tribune*, January 2, 1936, 4; "Cosmic Clearance," *TIME*, January 13, 1936, 28–33.

2. "New Cosmic Ray Tests Planned by Dr. Compton," *Chicago Tribune*, January 15, 1936, 14; "Dr. Compton Leaves Today for Ray Tests," *Chicago Tribune*, January 17, 1936, 16; "Compton Hails Cosmic Ray Tests Aboard Ship," *Chicago Tribune*, March 21, 1936, 8; "Compton Awaits Cosmic Ray Data from Steamship," *Chicago Tribune*, March 31, 1936, 11.

3. Arthur Compton to Curtis G. Benjamin, January 11, 1949, AHC Papers, series 3, box 11, folder 14-02; drafts in AHC Papers, series 2, box 5, folder 7; series 2, box 6, folder 7.

4. AHC Papers, series 2, box 5, folder 6.

5. "Ray Theory Supported," *Los Angeles Times*, May 20, 1936, 11.

6. "What Are Cosmic Rays?," *New York Times*, May 3, 1936.

7. William S. Barton, "Our Expanding Universe," *Los Angeles Times*, June 7, 1936, 119.

8. Barton, "Our Expanding Universe," 119.

9. Barton, 119.

10. Arthur Compton to Robert Millikan, November 12, 1936, RAM Papers, 21.19.

11. "'Precision' Surveys of Cosmic Ray Intensity (draft with annotations)," RAM Papers, 21.19.

12. Jacob Clay to Arthur Compton, September 18, 1936, AHC Papers, series 2, box 5, folder 7.

13. Jacob Clay addendum to Arthur Compton paper, RAM Papers, 21.19.

14. Robert Millikan to Arthur Compton, December 4, 1936, RAM Papers, 21.19.

15. John Tate to Robert Millikan, December 2, 1936, RAM Papers, 21.19.

16. Arthur Compton to John Tate, November 21, 1936, AHC Papers, series 2, box 5, folder 7-02.

17. Robert Millikan to John Tate, December 7, 1936, AHC Papers, series 2, box 5, folder 7-02.

18. Arthur Compton to Robert Millikan, December 12, 1936, AHC Papers, series 2, box 5, folder 7-02.

19. Jacob Clay to Arthur Compton, December 15, 1936, AHC Papers, series 2, box 5, folder 7-02.

20. Arthur Compton to John Tate, January 8, 1937, AHC Papers, series 2, box 5, folder 7-02.

21. Robert Millikan to Arthur Compton, January 19, 1937, RAM Papers, 21.20.

22. De Maria and Russo, "Cosmic Ray Romancing," 258.

23. Robert Millikan to Arthur Compton, January 19, 1937, AHC Papers, series 2, box 5, folder 7-02.

24. Jacob Clay to Arthur Compton, January 27, 1937, AHC Papers, series 2, box 5, folder 7-02.

25. Robert Millikan to John Tate, January 28, 1937, RAM Papers, 21.20.

26. Millikan to Tate, January 28, 1937.

27. Robert Millikan to Arthur Compton, February 1, 1937, AHC papers, series 2, box 5, folder 7-02.

28. Arthur Compton to Robert Millikan, March 6, 1937, RAM Papers, 21.20.

29. John Tate to Robert Millikan, March 11, 1937, RAM Papers, 21.20.

30. Robert Millikan to John Tate, March 26, 1937, RAM Papers, 21.20.

31. Arthur Compton to Robert Millikan, April 1, 1937, AHC papers, series 2, box 5, folder 7-02.

32. Arthur Compton to Robert Millikan, August 23, 1937, RAM Papers, 21.20.

33. Robert Millikan to Arthur Compton, October 13, 1937, RAM Papers, 21.20.

34. De Maria and Russo, "Cosmic Ray Romancing," 260.

35. Kargon, *Rise of Robert Millikan*, 161.

36. Robert Millikan to Arthur Sulzberger, December 13, 1937, RAM Papers, 41.33; emphasis in original.

37. Arthur Sulzberger to Robert Millikan, December 16, 1937, RAM Papers, 41.33.

38. Jacob Clay to Arthur Compton, January 3, 1938, AHC papers, series 2, box 5, folder 7-02.

39. Arthur Compton to John Tate, January 17, 1938, ibid.

40. Jacob Clay, "Millikan About the Latitude Effect of Cosmic Rays," undated, AHC papers, series 2, box 6, folder 2.

CHAPTER 18

1. William Fulton, "New York Fair Opens in Burst of Hullabaloo," *Chicago Tribune*, May 1, 1939, 4.

2. "Einstein in New Triumph," *New York Times*, May 1, 1939.

3. "To Illuminate N.Y. World's Fair by Cosmic Ray," *Boston Globe*, March 20, 1939, 13; "Cosmic Rays Will Open Fair," *Los Angeles Times*, April 10, 1939, 24; "Cosmic Rays Start Brilliant Display," *New York Times*, May 1, 1939; "Cosmic Ray Trap Sets Alight New York Fair," *Popular Mechanics*, July 1939, 26.

4. Kargon, *Rise of Robert Millikan*, 166–168.

5. Marcia Bartusiak, *The Day We Found the Universe* (New York: Vintage Books, 2009), 149–150.

6. Marcel Shein to Arthur Compton, August 26, 1943, AHC Papers, series 3, box 3, folder 31.

7. Eric Nordern, "*Playboy* Interview: Stanley Kubrick," September 1968, in *Stanley Kubrick: Interviews* (Jackson: University Press of Mississippi, 2001).

POSTLUDE

1. Rossi, *Cosmic Rays*, 219.

BIBLIOGRAPHY

All other sources are listed in the notes.

Auger, Pierre. *What Are Cosmic Rays?* Chicago: University of Chicago Press, 1945.

Bartusiak, Marcia, ed. *Archives of the Universe: 100 Discoveries That Transformed Our Understanding of the Cosmos*. New York: Vintage Books, 2004.

Bartusiak, Marcia. *The Day We Found the Universe*. New York: Vintage Books, 2009.

Bertolotti, Mario. *Celestial Messengers: The Story of a Scientific Adventure*. Berlin: Springer-Verlag, 2013.

Brooks, Michael. *Free Radicals: The Secret Anarchy of Science*. New York: The Overlook Press, 2012.

Brown, Laurie M., and Lillian Hoddeson, eds. *The Birth of Particle Physics*. Cambridge: Cambridge University Press, 1983.

Clay, Roger, and Bruce Dawson. *Cosmic Bullets: High Energy Particles in Astrophysics*. Australia: Helix Books, 1997.

Compton, Arthur Holly. *The Cosmos of Arthur Holly Compton*. Edited by Marjorie Johnston. New York: Alfred A. Knopf, 1967.

Compton, Arthur Holly. *The Human Meaning of Science*. Chapel Hill: University of North Carolina Press, 1940.

DeVorkin, David H. *Race to the Stratosphere: Manned Scientific Ballooning in America*. New York: Springer-Verlag, 1989.

Dornan, Lev I., and Irina V. Dornan. *Cosmic Ray History*. New York: Nova Publishers, 2014.

Falkenburg, Brigitte, and Wolfgang Rhode, eds. *From Ultra Rays to Astroparticles: A Historical Introduction to Astroparticle Physics*. Dordrecht: Springer, 2012.

Friedlander, Michael W. *A Thin Cosmic Rain: Particles from Outer Space*. Cambridge, MA: Harvard University Press, 1989.

Gaisser, Thomas K., Ralph Engel, and Elisa Resconi. *Cosmic Rays and Particle Physics*. Cambridge: Cambridge University Press, 2016.

Galison, Peter. *How Experiments End*. Chicago: University of Chicago Press, 1987.

Galison, Peter. *Image and Logic: A Material Culture of Microphysics*. Chicago: University of Chicago Press, 1997.

Goodstein, Judith R. *Millikan's School: A History of the California Institute of Technology*. New York: W. W. Norton & Co., 1991.

Kargon, Robert H. *The Rise of Robert Millikan: Portrait of a Life in American Science*. Ithaca, NY: Cornell University Press, 1982.

Kevles, Daniel J. *The Physicists: The History of a Scientific Community in Modern America*. Cambridge, MA: Harvard University Press, 1995.

Lemon, Harvey Brace. *Cosmic Rays Thus Far*. New York: W. W. Norton & Co., 1936.

Millikan, Robert Andrews. *The Autobiography of Robert A. Millikan*. New York: Prentice-Hall, 1950.

Millikan, Robert Andrews. *The Electron: Its Isolation and Measurement and the Determination of Some of Its Properties*. Edited with an introduction by Jesse W. M. DuMond. Chicago: University of Chicago Press, [1917] 1968.

Millikan, Robert Andrews. *Science and the New Civilization*. New York: Charles Scribner's Sons, 1930.

Pais, Abraham. *Inward Bound: Of Matter and Forces in the Physical World*. Oxford: Oxford University Press, 1986.

Rosen, Stephen, ed. *Selected Papers on Cosmic Ray Origin Theories*. New York: Dover, 1969.

Rossi, Bruno. *Cosmic Rays*. New York: McGraw-Hill, 1964.

Ryan, Craig. *The Pre-Astronauts: Manned Ballooning on the Threshold of Space*. Annapolis, MD: Naval Institute Press, 1995.

Sekido, Yataro, and Harry Elliot, eds. *Early History of Cosmic Ray Studies: Personal Reminiscences with Old Photographs*. Dordrecht, Holland: D. Reidel Publishing Company, 1985.

Steber, George R. *Invisible Visitors from Outer Space: Cosmic Rays, The Story from the Beginning to the Present*. Self-published, 2021.

Swann, W. F. G. *The Nature of Cosmic Rays*. Cambridge, MA: Sky Publishing Corporation, 1945.

Swann, W. F. G. *The Story of Cosmic Rays*. Washington, DC: Smithsonian Institution, 1957.

INDEX

Page numbers in italics refer to figures.